Advances in Business, Operations, and Product Analytics

Advances in Business, Operations, and Product Analytics

Cutting Edge Cases from Finance to Manufacturing to Healthcare

Matthew J. Drake

Publisher: Paul Boger
Editor-in-Chief: Amy Neidlinger
Executive Editor: Jeanne Glasser Levine
Editorial Assistant: Kristen Watterson
Cover Designer: Chuti Prasertsith
Managing Editor: Kristy Hart
Senior Project Editor: Betsy Gratner
Copy Editor: Karen Annett
Proofreader: Debbie Williams
Indexer: Lisa Stumpf
Senior Compositor: Gloria Schurick
Manufacturing Buyer: Dan Uhrig

©2016 by Matthew J. Drake
Published by Pearson Education, Inc.
Old Tappan, New Jersey 07675

For information about buying this title in bulk quantities, or for special sales opportunities (which may include electronic versions; custom cover designs; and content particular to your business, training goals, marketing focus, or branding interests), please contact our corporate sales department at corpsales@pearsoned.com or (800) 382-3419.

For government sales inquiries, please contact governmentsales@pearsoned.com.

For questions about sales outside the U.S., please contact international@pearsoned.com.

Company and product names mentioned herein are the trademarks or registered trademarks of their respective owners.

All rights reserved. No part of this book may be reproduced, in any form or by any means, without permission in writing from the publisher.

Printed in the United States of America

First Printing September 2015

ISBN-10: 0-13-396370-5
ISBN-13: 978-0-13-396370-0

Pearson Education LTD.
Pearson Education Australia PTY, Limited
Pearson Education Singapore, Pte. Ltd.
Pearson Education Asia, Ltd.
Pearson Education Canada, Ltd.
Pearson Educación de Mexico, S.A. de C.V.
Pearson Education—Japan
Pearson Education Malaysia, Pte. Ltd.

Library of Congress Control Number: 2015943206

*For my wife, Nicole,
and my daughters, Noelle and Maia.
You are the inspiration
for everything that I accomplish.*

Contents

Preface xx

Part I General Business Analytics 1

Chapter 1 VidoCo Demand Forecast Information Sharing 3
  Background 3
  VidoCo and TechnoMart 4
  Study Questions 9
  Endnotes 9
  Exhibits 10

Chapter 2 Container Returns at Pasadena Water Solutions 15
  PWS's People Planet Profit (P3) Sustainability Initiative .. 16
  PWS's Operations 18
  The Water Systems Shipping Containers 19
  The Container Supply Chain 21
  Mike's Options 22
  Exhibit 26

Chapter 3 Developing a Business Model to Improve the Energy Sustainability of Existing Buildings 27
  Introduction 27
  Improving Energy Efficiency at Boğaziçi Sarıtepe Campus 29
  Energy Efficiency Measures Identified for Boğaziçi Sarıtepe Campus 30
  Selecting the Energy Efficiency Measures Optimally for a Given Budget 32
  Selecting Energy Efficiency Measures in Multiple Time Periods 33
  Offering Energy Efficiency Measures as a Service: ESCO Business Model 34
  Investing in Energy Efficiency Measures as an Alternative Source of Energy 35
  Study Questions 36
  Endnote 37
  Exhibits 38

Chapter 4	Contracting for LTL Services at Hankey Industries	43
	Hankey Industries' Logistics Operations	43
	Contracting for LTL Services	45
	LTL Rate Quotes	46
	Selecting Contracts for LTL Services	48
	Endnotes	49
	Exhibits	49
Chapter 5	Optimal Product Bundling at Point and Shoot Camera Shop	61
	Company Background	61
	Product Bundling	62
	Market Research	64
	Developing an Optimal Bundling Strategy	66
	Exhibits	66
Chapter 6	Intermodal Routing for an Ocean Container	69
	Company Background	70
	Intermodal Routing Options	71
	Route Selection Decision	72
	Exhibits	73

Part II Analytics in the Service and Utility Industries 75

Chapter 7	Lolly's Restaurant	77
	Study Questions	80
	Endnote	80
	Exhibits	81
Chapter 8	Strategies for Managing Service Delivery Gaps and Service Recovery	83
	Background on the Hotel Industry in India	84
	Service Failure	86
	Service Recovery	89
	Service Process Improvement	90
	Conclusion	91
	Endnotes	91
	Additional Readings	92
	Exhibit	93

Chapter 9	EverClean Energy, Inc.: Wind Energy Versus Natural Gas...95	
	Introduction 95	
	Company Background 96	
	Wind Energy....................................... 97	
	Drivers of Wind Energy Costs 98	
	The Declining Cost of Wind Energy 99	
	Factors Affecting the Cost to Generate Electricity from Natural Gas 100	
	Conclusion....................................... 102	
	Acknowledgments................................ 103	
	Endnotes 103	
	Exhibits 106	
Chapter 10	Slotting Pharmaceuticals in an Automated Dispensing Cabinet 111	
	Southeast Ohio Women's Hospital 111	
	Medication Profile 112	
	Strategy for Stocking the ADC.................... 113	
	ADC Benefit Analysis............................. 114	
	Endnotes 116	
	Exhibits 117	
Part III	**Analytics in Accounting and Finance 121**	
Chapter 11	Tax Ramifications of S Corporation Shareholder Termination or Change of Ownership Interest 123	
	Introduction 123	
	Decision Scenarios and Discussion 125	
	Conclusion....................................... 129	
	References....................................... 129	
Chapter 12	Charleston Rigging 131	
	History of Charleston Rigging 131	
	The Opportunity................................. 134	
	The Marketing Study 135	
	Endnote... 136	
	Exhibits 137	

Chapter 13 Equipment Purchase and Replacement Strategy
at the Fayette China Company 139
  Company Background 140
  Equipment Replacement Strategy. 141
  Equipment Purchase Decision. 142
  Exhibits 143

Part IV Analytics in the Public Sector 145

Chapter 14 Using Regression to Improve Parole Board Decisions . 147
  Study Questions 150
  Closing Comments 151
  Exhibit 151

Chapter 15 Redesigning Pittsburgh Port Authority's Bus Transit
System. 153
  Introduction 153
  Pre-analysis 155
  Model 156
  Data for Case Analysis 158
  Study Questions 159
  Endnote. 161
  Exhibits 162

Part V Analytics in Management and Ethical Decision
Making 165

Chapter 16 The Bloodgate Affair: A Case of Breaking Rules and
Breaching Trust? 167
  Introduction 167
  The Incident 170
  The Immediate Aftermath 171
  The Outcome of the ERC Disciplinary Panel Hearing . . 172
  Responses to the Verdict 173
  The Delayed Aftermath 174
  Epilogue 177
  The Task 179
  Study Questions 180

Endnotes................................ 182
Additional Reading....................... 189
Exhibit................................. 190

Chapter 17 Trouble on the Thames: Event Disruption,
Public Protest, or Public Disorder 191

Introduction 191
The Act of Protest at the Oxford-Cambridge Boat Race . 192
The Protester 193
The Immediate Aftermath 194
The Reaction........................... 194
Epilogue 198
The Task 199
Study Questions 200
Endnotes.............................. 202
Additional Reading..................... 207
Exhibit................................ 208

Index. 209

Note: Data Files Available Online

Data files corresponding to the case studies can be found on the book website, www.ftpress.com/title/9780133963700. Click the Downloads tab to access them.

Acknowledgments

I am forever grateful for the efforts of all of the contributors to this book. Many of them have been friends and colleagues for a long time, but I met some others for the first time through working on this project. I look forward to many future opportunities to collaborate with them. This book would not have become a reality without the contributors' willingness to share their hard work with me and the readers of this book. I am also indebted to Jeanne Glasser Levine, Executive Editor at Pearson, whose guidance and advice was instrumental throughout the publication process. This is the second project that I've worked on with Jeanne, and both occasions have been extremely smooth and pleasant experiences. It is extremely helpful to have such a strong advocate for my work. I sincerely hope that our collaboration will continue in the future on subsequent projects.

About the Author

Matthew J. Drake, Ph.D., CFPIM, is an Associate Professor of Supply Chain Management in the Palumbo-Donahue School of Business at Duquesne University. Dr. Drake primarily teaches analytical courses in the Supply Chain Management program. He holds a B.S. in Business Administration from Duquesne University and an M.S. and Ph.D. in Industrial Engineering from the Georgia Institute of Technology. His first book, *Global Supply Chain Management*, was published by Business Expert Press in 2012. His case book, *The Applied Business Analytics Casebook*, was published by FT Press in 2014. Dr. Drake's research has been published in a number of leading journals, including *Naval Research Logistics*, the *European Journal of Operational Research*, *Omega*, the *International Journal of Production Economics*, *OR Spectrum*, the *Journal of Business Ethics*, and *Science and Engineering Ethics*. Several of his previous cases and teaching materials have been published in *INFORMS Transactions on Education* and *Spreadsheets in Education*.

Dr. Drake lives in suburban Pittsburgh, Pennsylvania, with his wife, Nicole, his daughters, Noelle and Maia, and his dog, Bismarck.

About the Contributors

Emre Çamlıbel is the CEO of the Soyak Group. He received his B.S. degree from Yıldız Technical University, his M.S. degree from MIT, and his Ph.D. in Civil Engineering from Boğaziçi University. His Ph.D. dissertation was on developing an optimization model toward energy efficiency for existing buildings.

Patrick Cellie is an environmental engineer specializing in sustainability. For almost ten years, he has worked and provided consulting services related to sustainability topics across many industries, including both manufacturing companies and service providers. He holds a B.S. in Environmental Engineering from University of Rome "La Sapienza" in Italy and an MBA from Duquesne University. He currently lives in Pittsburgh, Pennsylvania.

Elizabeth Conner, CPA, is a member of the faculty at the University of Colorado at Denver. Ms. Conner teaches courses in taxation and financial accounting; she is also a practicing CPA. She has written articles and case studies that focus on accounting and income tax issues. Her most recent case study is being offered on an intermediate accounting textbook's website and her most recent article was published in *The Tax Adviser*.

Duygu Dagli is an industrial engineer specializing in operations management and business analytics. She has worked and provided consulting services in the manufacturing, technology, healthcare, and tourism industries. She currently works as the revenue manager of Turnkey Vacation Rentals. She holds a B.S. in Industrial Engineering from Bilkent University, an MBA from Middle East Technical University, and an M.Sc. in Supply Chain Management from University of Texas at Dallas. She lives in Austin, Texas.

John Davies is Professor of Management and Associate Dean (International and Executive Education) at the Victoria Business School, Wellington, New Zealand. He graduated from the University of Wales and the University of Lancaster with a background in operational research and has developed his research interests primarily within the fields of the decision sciences and systems methodologies. He has published widely in leading academic journals spanning the decision sciences, technology management, systems, and sports management.

Steven Harrod received his Ph.D. from the University of Cincinnati and is an Associate Professor at the Department of Transport, Technical University of Denmark. He is coordinator of the Railway Technology study line and the Public Transport research group. Dr. Harrod supervises courses in transportation management and optimization.

Fatma Kılınç-Karzan is an Assistant Professor of Operations Research at the Tepper School of Business at Carnegie Mellon University. She received her Ph.D. from the Georgia Institute of Technology. Her research interests lie in large-scale optimization methodologies, especially related to handling uncertainty and their applications in business analytics and machine learning. She has several articles in journals including *Annals of Statistics, Interfaces, Mathematical Programming, Operations Research Letters, SIAM Journal on Matrix Analysis and Algebra*, and *International Journal of Production Research*.

Ersin Körpeoglu is a Ph.D. Candidate in Operations Management at the Tepper School of Business at Carnegie Mellon University. His research interests lie in game-theoretical analysis of innovation and supply chain management problems, as well as business analytics and production scheduling applications. His work has appeared in several journals, including *Interfaces, the European Journal of Operational Research*, and *Computers and Operations Research*.

Doug Laufer, CPA, CFE, is a Professor of Accounting at Metropolitan University of Denver. Dr. Laufer's research focus is on accounting education, teaching pedagogy, and occupational fraud. His publications include authored or coauthored articles appearing in the *Journal of Forensic Studies in Accounting and Business, American Journal of Economics and Business Administration, The Tax Adviser* and *Advances in Accounting Education.*

Thomas McCue is an Associate Professor of Finance at Duquesne University in Pittsburgh, Pennsylvania. He has been teaching with cases for more than 35 years. He currently teaches the undergraduate capstone case course that is required for all finance students. He received his Ph.D. from the University of North Carolina at Chapel Hill. He has written numerous research articles in finance and real estate.

Emre N. Otay is an Associate Professor of Civil Engineering at Boğaziçi University. He has a B.S. degree in Civil Engineering from Boğaziçi University; a Dipl. Ing. degree in Civil Engineering from the Technical University, Braunschweig; and a Ph.D. degree in Coastal and Oceanographic Engineering from the University of Florida. His research interests include renewable energy, energy efficiency of existing buildings, and stochastic analysis and modeling. He is currently the Coordinator of the Boğaziçi University Sarıtepe Campus and the Boğaziçi University Wind Power Plant project.

Özalp Özer is the Ashbel Smith Professor of Management Science at the Jindal School of Management of the University of Texas at Dallas. Professor Özer's areas of specialty include end-to-end management and design of global supply chains, capacity/inventory planning, logistics design, incentive/contract design, retail and pricing management. He received his Ph.D. and M.S. degrees in Operations Research from Columbia University.

Pedro M. Reyes received his Ph.D. in Operations Management, an MBA in Operations Management, an M.S. in Information Systems, and a B.S. in Mathematics from the University of Texas at Arlington (UTA). He is an Associate Professor in the Hankamer School of Business at Baylor University and is the Director for the Center of Excellence in Supply Chain Management. Dr. Reyes currently teaches the courses in operations management and global supply chain management. His research interests consist of global supply chain and logistics operations planning and control systems, radio frequency identification humanitarian operations, and crisis management. Dr. Reyes is the author of *RFID in the Supply Chain* (published by McGraw Hill in 2011) and has published numerous journal articles. In 2005 he received the Hankamer School of Business Young Researcher Award (2005), and in 2002 he was recognized by the UTA College of Business as a *Lawrence Schkade Research Fellow*. He was also a winner of the 2014 Decision Sciences Institute Best Case Studies Award Competition.

Chris Roethlein is a Professor in the Management Department at Bryant University where he teaches courses in operations management and supply chain management. He has a Ph.D. in Management Science and Information Systems from the University of Rhode Island, and his research interests include quality and communication within a supply chain, strategic initiatives through alignment of supply chain goals, collaborative relationships, and leadership excellence. He has published in numerous journals, and he was a winner of the 2011 and 2014 Best Case Studies Award Competition sponsored by the Decision Sciences Institute.

Wendy Swenson Roth is an Associate Professor in the Managerial Sciences Department at Georgia State University in Atlanta, GA. She received her Ph.D. from the University of Tennessee. Her research interests include the use of technology in the classroom to improve student outcomes. Dr. Roth's courses include business intelligence, analysis, and modeling.

Barış Tan is a Professor of Operations Management at Koç University. He received a BS degree in Electrical and Electronics Engineering from Boğaziçi University, a master's degree in Industrial and Systems Engineering, a manufacturing systems engineering degree, and a doctor of philosophy degree in operations research from the University of Florida. His current research interests are in developing business models for energy efficiency investments, stochastic modeling of operations, performance evaluation and control of manufacturing systems, and cooperation, subcontracting, and outsourcing in supply chain management.

Ajith K. Thomas holds an M.B.A in Marketing and Ph.D. in Management Sciences, with his first professional degree in Hotel Management. He has held senior positions in corporate marketing with multinational hospitality companies for about 10 years and has been an Associate Professor at Saintgits Institute of Management, Kottayam, India, for about 6 years. He is also a visiting academic at Duquesne University in Pittsburgh, PA. His areas of research interest include services marketing and consumer behavior.

John K. Visich is a Professor in the Management Department at Bryant University where he teaches courses in operations management, supply chain management, and international business. He has a Ph.D. in Operations Management from the University of Houston, and his research interests are in supply chain management, radio frequency identification, and corporate social responsibility. He has published in numerous journals, and in 2011 and 2014 he won the Decision Sciences Institute Best Case Studies Award Competition. He is also a four-time recipient of the student-selected Outstanding MBA Professor Award at Bryant University, and in 1999 he received the Melcher Award for Excellence in Teaching by a Doctoral Candidate from the C.T. Bauer College of Business of the University of Houston.

Yahya Yavuz received his B.A. degrees in Mathematics and Business Administration from İzmir Ekonomi University and his M.S. in Industrial Engineering from Koç University. After working at North Atlantic Consulting Group, he is currently CRM Segmentation and Service Modelling Assistant Manager at Akbank.

Yanchong Zheng is the Sloan School Career Development Professor and Assistant Professor of Operations Management in the Sloan School of Management at the Massachusetts Institute of Technology. Professor Zheng's expertise is in the areas of behavioral operations management, sustainable supply chains, consumer behavior and retail operations, and behavioral and experimental economics. She received her B.Eng. and M.Eng. degrees from Tsinghua University, China, and her Ph.D. degree from Stanford University.

Preface

The field of business analytics continues to gain momentum as more organizations begin to emphasize its importance in improving the effectiveness of the decision-making process. Thomas H. Davenport perhaps did the most to thrust the limelight on the field with his seminal 2006 *Harvard Business Review* article, "Competing on Analytics," which he and coauthor Jeanne G. Harris later expanded into a book with the same title. In the decade since Davenport's initial publication, business schools around the world have rushed to ensure that their curricula reflected the analytics trend. Many programs have even established "business analytics" concentrations or minors that can enhance the value of students' degrees.

The interesting thing about this recent analytics wave is that business analytics itself really is not that new of a concept. A general definition of business analytics is "the scientific process of transforming data into insight for better decision making."[1] Organizations have been using data analysis to inform their decision making for decades. The more recent development and proliferation of desktop technology applications for analytics have expanded the accessibility of these solutions to a wider range of business professionals instead of relegating them to those experts in computer coding and programming as was the case in the past. Business schools no longer have to reserve coverage of these topics for students who have taken a suite of programming courses because much of the analysis can be performed in a spreadsheet, possibly utilizing add-in tools with point-and-click interfaces.

With more business school programs offering courses in business analytics and incorporating analytics material into their existing courses, there is a need for additional offerings in the library of educational materials. This book is designed to serve dedicated analytics courses as well as instructors in other functional areas who want to introduce analytics into the coverage of their core material. Cases are an especially effective way to teach analytics because they place the students in a simulated role as a decision maker in an organization. The cases often provide enough detail that the students must identify the information that is relevant to the analysis at hand rather than neatly organizing the information that they will need as textbook problems usually do. These additional details also provide topics for follow-up discussion in the course beyond the original analysis. Instructors can emphasize not only the mechanics of the technical analysis but also the way that the analytical results can be used to help a manager make better decisions.

Several vast case libraries are maintained throughout the world, which allow instructors to identify materials that complement their course content. Many of these cases are written at a level that is only appropriate for graduate students, and it can be challenging to find cases that are appropriate for undergraduates. To address that gap, this book mainly contains cases that could be used effectively at either the undergraduate or the introductory graduate level. They are also of varying length, with some being relatively short for use in a 30-minute class discussion and longer ones that are more appropriate for an out-of-class assignment and subsequent wrap-up discussion.

The cases in this collection are grouped by the business or industry application highlighted in the case. This structure allows instructors of courses in various business functions to identify quickly the cases that are most appropriate for their courses. The degree and sophistication of analysis required varies greatly from case to case, with some cases demanding extensive quantitative modeling and analysis and other cases necessitating a more qualitative approach. Part I includes six "general" business analytics cases that apply to business functions such as demand planning, logistics, and sustainability. Part II contains four cases set in organizations within the service and utility industries. Part III includes three cases that require students to apply analytics to accounting and finance decision environments. Part IV contains two cases from the public sector to provide a government and nonprofit decision-making perspective. Part V provides two cases that utilize analytics to aid the development of ethical decisions.

> **Note: Data Files Available Online**
>
> Data files corresponding to the case studies can be found on the book website, www.ftpress.com/title/9780133963700. Click the Downloads tab to access them.

It is my hope that the cases in this collection expose students to the opportunity that exists to apply business analytics to improve decision making in organizations in a variety of industries. Students equipped with an effective set of analytical skills and techniques will be valuable contributors to their companies and organizations as a result of their ability to make thoughtful, reasoned decisions informed by data analysis. The broad applicability of these analytical skills will serve the students well regardless of where their careers may take them in the future.

Matthew J. Drake
Pittsburgh, Pennsylvania, USA
May 2015

Endnote

1. Data source: http://www.informs.org/About-INFORMS/What-is-Analytics

Part I
General Business Analytics

1 VidoCo Demand Forecast Information Sharing 3

2 Container Returns at Pasadena Water Solutions 15

3 Developing a Business Model to Improve the Energy Sustainability of Existing Buildings 27

4 Contracting for LTL Services at Hankey Industries 43

5 Optimal Product Bundling at Point and Shoot Camera Shop 61

6 Intermodal Routing for an Ocean Container 69

1

VidoCo Demand Forecast Information Sharing[1]

Duygu Dagli, Turnkey Vacation Rentals
Özalp Özer, University of Texas at Dallas
Yanchong Zheng, Massachusetts Institute of Technology

Background

Demand forecasting and forecast information affect key operational and strategic decisions in managing global supply chains. Companies often face long lead times in procuring materials and production. Hence, they rely on demand forecasts to determine production capacity and inventory stocking levels. As a result, obtaining accurate demand forecasts is critical to meet the final market demand. Typically, when a company is closer to the end customer (e.g., consumer-facing companies such as Apple and Nike, and retailers such as Walmart and Best Buy), the company can gain more information about the market demand and hence is able to generate more accurate demand forecasts. Therefore, an upstream supplier usually relies on demand forecasts provided by a downstream company to plan its production.

Realizing the importance of demand forecast sharing in a supply chain, companies such as General Motors, Procter & Gamble, and Neiman Marcus have invested heavily in deploying information management systems within their global supply chains to better coordinate with upstream suppliers. The development of initiatives such as electronic data interchange (EDI) and collaborative planning, forecasting, and replenishment (CPFR) has offered companies technological solutions for sharing demand forecast information. However, the effectiveness of such initiatives remains uncertain because these technologies facilitate information sharing but do not necessarily ensure the shared information to be credible or accurate. Companies continue to face critical challenges to enable effective forecast information sharing with their supply chain partners.

VidoCo and TechnoMart

In June 2012, VidoCo, a medium-sized company specializing in webcams, reached a distribution/sales agreement with TechnoMart, one of the largest consumer electronics retailers in the United States, to sell high-definition (HD) webcams. The retail sales would commence in July 2012. Five months after the contract was signed, VidoCo started to face a dramatic challenge of determining the correct production volume to fulfill the retail demand in upcoming months.

VidoCo was established in China in 2002 as a medium-sized company that supplies webcams for local laptop computer manufacturers. VidoCo's focus on research and development and high-quality production, coupled with the rapid growth of the Internet during the mid-2000s (Exhibit 1.1), helped it to become a major supplier of webcams in China. In 2009, VidoCo announced its strategic plan to introduce stand-alone HD Bluetooth webcams into its product lines and expand to the global market, in particular North America. The plan had the vision of gaining higher profit margins and reaching a wider customer base. VidoCo introduced its first HD Bluetooth webcam models in March 2012. Its HD product line has 12 different webcam models that differ from each other in terms of the product design.

Apart from design, the same components and a similar bill of materials are used to produce these webcams. VidoCo's research and development (R&D) team follows the technological developments closely and recommends upgrades. Hence, VidoCo has to change the lens or the electronic components of all its products once a new technology emerges. On average, the product life cycle of a particular model is 18 months. VidoCo's product samples were demonstrated in various technology fairs, including the one that led to the contract with TechnoMart.

TechnoMart is a consumer electronics retailer with more than 640 stores in the United States. Its brick-and-mortar retail stores and online store generated an annual profit of $5.23 billion in 2012. Its merchandise includes personal computers, computer accessories, digital cameras, video cameras, mobile phones, and software. TechnoMart has 85 suppliers, most of which are located in East Asia. As part of the company's policy, TechnoMart looks for innovative and high-quality products with low costs. Jason McDonell, a procurement manager at TechnoMart, took notice of VidoCo's webcams during his visit at the Computex Taipei Fair in May 2012. During a monthly meeting at TechnoMart's headquarters in Dallas, Jason recommended VidoCo's webcams to the TechnoMart sales manager. Eventually, the company decided to sell VidoCo's webcams through its stores nationwide.

VidoCo received an invitation from TechnoMart a few weeks later. During the meeting, VidoCo presented its products and the negotiation process was initiated. TechnoMart uses a single per-unit price point to simplify the procurement process. Under this single-price contract, TechnoMart will pay VidoCo a fixed wholesale price for each unit of webcams it purchases regardless of the total purchase quantity. To start production quickly and catch the upcoming holiday season, the Sales Department at VidoCo accepted the contract promptly and requested a manufacturer's suggested retail price (MSRP) to be $60 because the company targets the high-end market with its high-quality products. Eventually, both firms agreed that TechnoMart would sell the webcam at a per-unit retail price of $75, and pay VidoCo a wholesale price of $45 for each unit in the final order. This outcome was a profitable deal for VidoCo because its production cost was $28 per unit. TechnoMart's representatives requested that they first conduct a market analysis for HD webcams before committing to the initial order quantity.

The U.S. consumer electronics retail industry is highly competitive. Even though specialized retailers, such as TechnoMart, dominate the consumer electronics sales (69%), grocery retailers and Internet retailers capture 20% of the market.[2] The majority of customers have low loyalty to brands or stores, and they are interested in innovative and attractively priced products. As a result, the retailers have been competing through deep discounts, bundling popular items with complementary products or services (such as installations or extended warranties), introducing new products into their portfolios, and matching their prices with the competitors'.

The Marketing and Sales (M&S) Department of TechnoMart estimated that the demand for the new webcams could be any number between 5,500 and 8,600 units per month, equally likely. Because TechnoMart cannot predict its competitors' pricing and advertising strategies for the upcoming months, the M&S Department estimated that there would be an unpredictable demand variation added to the estimate. This additional demand variation is likely to be uniformly distributed between −850 and 850. Given these estimates, the lowest and highest possible demand for the webcam would be 4,650 and 9,450 per month.

Jason McDonell and Sheng Fang, the brand manager of VidoCo, met at TechnoMart's headquarters in Dallas, Texas, in June 2012. During the meeting, Jason presented their first demand estimate and explained TechnoMart's sales strategy for VidoCo's webcams. Each month, TechnoMart offers special discounts for particular desktops, laptops, and netbooks. Which computers will be discounted is determined based on TechnoMart's inventory levels for the particular products and its agreements with suppliers. Because demand shifts to the computers on discount, TechnoMart would offer complementary products, such as webcams, speakers, keyboards, and mouses with the computers. TechnoMart would select the webcam that matches the discounted computer and inform VidoCo in advance of the selected webcam that it would order. Given the wide variety of HD webcam models and their short life cycles, Jason commented that a particular model would be selected only once. Jason also mentioned that TechnoMart is particularly careful about supply lead times and inventory levels. The company maintains high product availability and avoids stockouts. To help its suppliers to better plan for production and shipments, TechnoMart's policy with its suppliers has been to communicate the company's demand forecasts for the upcoming three months on a monthly and rolling horizon basis. TechnoMart places a final order at the beginning of each month, and this final order must be fulfilled within 20 days. Jason pointed out that the company would only purchase the quantity indicated in the final order.[3] TechnoMart also could not guarantee that the quantity in the final order would be the same as the forecasts sent earlier because market demand is variable. Being optimistic about the relationship and eager to catch the holiday season, Sheng, on behalf of VidoCo, accepted these terms. Shortly after Sheng returned to China, TechnoMart sent its first set of demand forecasts and the final order for July 2012. VidoCo fulfilled the first order from its inventory and started production for the coming months.

Five months later, Assistant Financial Manager Bo-Liu Ran of VidoCo noticed that inventory costs were rising. To resolve the issue, he organized a meeting with Brand Manager Sheng Fang, Assistant Supply Chain Manager Tao Li, and Assistant Production Manager Shi Dong.

Bo-Liu started the meeting by voicing his concerns about inventory costs. Shi commented that VidoCo was accumulating raw materials and inventories of webcams because of frequent changes in the orders placed by the Sales Department. Sheng Fang, the only representative from the Sales Department, presented the demand forecasts and orders received from TechnoMart (Exhibits 1.2 and 1.3). After investigating the data, all four managers noticed that the forecasts received have been substantially higher than the quantities in the final orders.

Shi came up with a suggestion: The forecasts received from TechnoMart were obviously misleading, so it might be better if VidoCo generated its own forecasts. Bo-Liu then suggested delaying the production until after receiving the final order. To respond, Tao provided the list of suppliers and the current supply lead times for all of the components (Exhibit 1.4). Shi added that VidoCo's assembly, quality control, and packaging processes took 5 hours and the plant had the capacity to produce at most 300 units per day. Sheng wondered whether VidoCo could change its suppliers to shorten the component supply lead times. Tao stated that the current suppliers satisfied the quality requirements at very reasonable prices. To do so, the suppliers have substantially reduced production and replenishment lead times. Requesting a further reduction in lead time may lead to higher procurement prices or lower production quality due to rush orders.

After an hour of discussion, Bao-Liu, Sheng, Tao, and Shi could not reach any solutions. They decided to end the meeting, individually think about possible solutions, and meet again in a week.

Study Questions

1. What are some possible reasons for TechnoMart to post optimistic forecasts (i.e., forecasts that are higher than final purchase quantities)?
2. Given TechnoMart's most recent forecasts for the three months from December 2012 to February 2013, how many units of HD 118-80, HD 127-A3, and HD 98-112 should VidoCo produce?
3. Evaluate the solutions proposed during the first meeting of the VidoCo managers. What are their pros and cons? Can you think of an alternative approach to address the problem?
4. How can VidoCo receive more reliable forecasts from TechnoMart?

Endnotes

1. The authorship is listed in alphabetical order. Ms. Dagli was a graduate student working under the supervision of Drs. Özer and Zheng while this case was written. This case was written for the purpose of class discussion, rather than to illustrate either effective or ineffective handling of a managerial situation.
2. Data source: Euromonitor International (December 2011). Consumer Electronics Global Overview: Growth Trends and Analysis. Web. Last accessed on 02/25/2013. http://www.euromonitor.com/consumer-electronics-global-overview-growth-trends-and-analysis/report.
3. TechnoMart's final order is equal to the actual demand for that month.

Exhibits

Exhibit 1.1 Internet Users as Percentage of Population

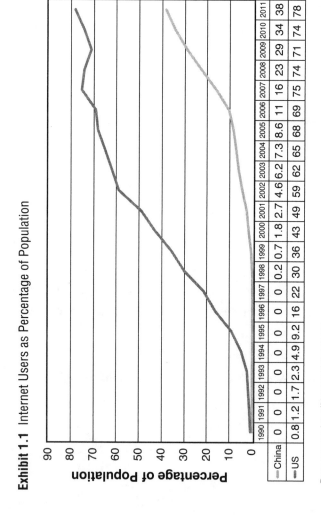

	1990	1991	1992	1993	1994	1995	1996	1997	1998	1999	2000	2001	2002	2003	2004	2005	2006	2007	2008	2009	2010	2011
China	0	0	0	0	0	0	0	0	0.2	0.7	1.8	2.7	4.6	6.2	7.3	8.6	11	16	23	29	34	38
US	0.8	1.2	1.7	2.3	4.9	9.2	16	22	30	36	43	49	59	62	65	68	69	75	74	71	74	78

Data source: http://data.worldbank.org

Exhibit 1.2 Forecasts and Orders Received from TechnoMart

Model	July 2012 HD 127-A3	August 2012 HD 122-B	September 2012 HD 192-B8	October 2012 HD 121-LP	November 2012 HD 410-F	December 2012 HD 118-80	January 2013 HD 127-A5	February 2013 HD 98-112
Final order	6,100	6,750	6,400	5,700	8,050			
Forecasts received in July		7,200	7,450	7,200				
Forecasts received in August			7,000	7,350	8,400			
Forecasts received in September				6,300	8,350	7,200		
Forecasts received in October					8,200	7,700	6,400	
Forecasts received in November						7,650	6,500	7,100

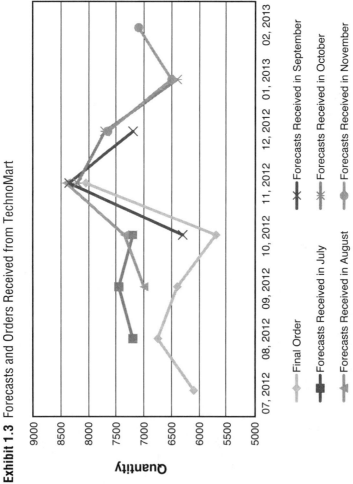

Exhibit 1.3 Forecasts and Orders Received from TechnoMart

Exhibit 1.4 List of Suppliers for HD Webcams

	Number of Items Required per Product	Supplier	Cost (per Unit)	Average Lead Time (Days)
A144 lens	1	Kunijo Co., Inc.	$5.00	38
CCD	1	Qualibit Co.	$2.00	14
Plastic frame	1	PlastyPlastronics Co.	$1.00	25
0.11-inch screws	4	Psiape Co.	$0.10	5
LED	1	Shiny Thingy Digitals Co.	$0.20	7
Bluetooth transceiver	1	Vexxfon Co.	$0.60	22
Microphone	1	Noisvape Co.	$0.12	16
Installation CD	1	Hihasu Co.	$0.10	8
Packaging material	1	Alphaberti Co., Inc.	$0.25	10

2

Container Returns at Pasadena Water Solutions

John K. Visich, Bryant University

Christopher J. Roethlein, Bryant University

Pedro M. Reyes, Baylor University

Mike Wolfe, manager of product packaging and distribution for Pasadena Water Solutions (PWS) in Pasadena, Texas, looked out his second-floor office and surveyed the landscape outside the manufacturing facility. It was early January 2012 and approximately 50 large, wooden containers used to ship water cleaning systems to customers had been returned to the facility as part of PWS's new People Planet Profit (P3) sustainability program. Prior to PWS's sustainability initiative, the disposal of a shipping container was handled by the customer. Now the containers were held in various storage locations around the facility and Mike was experiencing an increasing volume of returned shipping containers. Disposal by incineration for energy or by dumping in a landfill were options Mike did not prefer to use. He had to find a solution that would recover monetary value from the containers and have a positive impact on his bottom line, as well as support PWS's P3 sustainability initiative. And, he had to do this soon because containers were arriving every week at the plant and he was running out of space to store them.

PWS's People Planet Profit (P3) Sustainability Initiative

Over the past several years, PWS had come under increasing pressure from a growing number of customers to become a more sustainable company. Many of these customers had been actively practicing and publicizing their sustainability programs for many years, and they were in good standing with their own customers and traditional stakeholders. However, nongovernmental organizations (NGOs) had become an increasingly vocal new stakeholder, and the NGOs had moved beyond environmental and social activism at the company level. They were now focused on the sustainability of the supply chain and were demanding a higher level of transparency along the entire supply chain. Because it was difficult for the NGOs to monitor companies upstream in the supply chain, they were expecting the purchasing company to set supplier sustainability standards and audit their suppliers for compliance. In essence, a greening of the supply chain was taking place, with the largest or most powerful member of the supply chain acting as the "green enforcer."

In order to meet the environmental and social challenges they were now facing, PWS embarked on a companywide sustainability program in the spring of 2011. Each division of the company conducted a review of its operations to identify where it could reduce the environmental impact and/or have a positive social impact on employees and the local community. All proposals were to be analyzed from a P3 perspective. From this review, the different divisions introduced a variety of projects and programs that would make PWS a more visibly sustainable company. Any idea that could not be fulfilled from a division's operating budget would require a cost-benefit analysis and full implementation plan in order to obtain funding at the corporate level.

Mike's Product Packaging and Distribution (PPD) group was part of the Manufacturing Division, which implemented several projects to support PWS's sustainability initiative. New lighting had been approved for the plant prior to the P3 sustainability initiative, and the traditional fluorescent lighting in the project specifications was replaced with low-energy usage LED (light-emitting diode) fixtures. In addition, motion detectors were installed to minimize the illumination of idle areas of the facility. All production lines went through a Kaizen blitz 5S program where efforts were made to recycle as many unusable items as possible. Mike reviewed the materials used to package products for shipping and, in collaboration with the plant manager, consolidated all plastic wrap used in the facility to one film that could be recycled. Mike also set a requirement that all wood used to construct shipping containers had to be certified by the Sustainable Forestry Initiative or by the Forest Stewardship Council. During the summer of 2011, the vice president of the Manufacturing Division announced to customers a policy of accepting all empty equipment containers with PWS incurring the return shipping cost. The vice president implemented this policy because it had a high "green" impact with PWS's customers and he could report the number of returns in the Manufacturing Division's contribution to PWS's soon-to-be-written annual corporate sustainability report.

When the first returned containers began to arrive at PWS's facility, nobody knew what to do with them, so they were stored in an unused location behind a stand of oak trees where they were obscured from view. They were out of sight and therefore out of mind until that storage space reached capacity in the mid-fall of 2011. After this saturation point, the containers were then placed in locations that were highly visible to employees, visitors, and people driving past the facility. What was a good green idea for the customer was now a growing problem for the Manufacturing Division. Because the containers had originated within the PPD group, the vice president of the Manufacturing Division decided at the end of the year that Mike and his team would be responsible for the disposal of the containers.

PWS's Operations

Pasadena Water Solutions originally began operations as Pasadena Oil Equipment (POE) at a facility in Pasadena, Texas. Pasadena is located just to the southeast of Houston, and the plant had easy access to highways and the Port of Houston. Manufacturing operations initiated in 1900 to produce pumps and other equipment used by the burgeoning oil industry. Over the years, POE developed a strong research and development expertise in product improvement through the application of near cutting-edge technology. In the 1950s, it slowly diversified its customer base beyond the oil and gas industry to include the chemical, heavy manufacturing, and power generation industries. Though all four of these industries had similar product needs, POE was still highly dependent on the oil and gas industry for a significant majority of sales. The new customers, on the other hand, enabled POE to attain higher economies of scale in operations, which gave POE a competitive advantage based on low cost. This increased profit, which POE invested in manufacturing sites and in new product development. One of the new products was a proprietary water cleaning system that enabled a facility to reuse wastewater from the production process. This closed-loop system offered two benefits: reduced cost to purchase water and reduced wastewater disposal cost. Due to concerns about the availability of water and the environmental issues around wastewater disposal, water systems contributed to a growing percentage of POE's sales and profits. In the mid-1990s, the oil and gas industry service provider sector began to consolidate as large firms such as Schlumberger and Baker Hughes acquired smaller firms in order to provide end-to-end services for the major energy-producing companies. These firms wanted to offer an array of energy-related services from exploration all the way through to extraction. This consolidation weakened POE's market position, and when it was approached as an acquisition target, the company negotiated a deal to sell off the industrial pump and equipment part of the business while retaining the water cleaning product line and the original facility in Pasadena. Due to the change in business focus, POE reincorporated as Pasadena Water Systems.

The Water Systems Shipping Containers

Water systems are composed of four main elements: pumps, pipes, water cleansing equipment, and filters. Each water system was custom designed for the unique water properties of the process and then shipped to customers in a specially designed container that held all the components of the water system. The container was constructed by Mike's woodshop employees and was made close to the promised shipping date in order to save space within the plant. Water system containers were approximately 8 feet wide by 8 feet high and ranged in length from 20 feet to 40 feet. The container length was contingent upon the volume of water that needed to be cleaned and 12 different container lengths were used. However, 75% of the containers were in the 36- to 40-foot range. The loaded containers were moved on regular flatbed trailers without special permits.

The container was composed of two components: a pallet and a cover. The pallet base was similar to pallets seen in everyday use except the size was much larger. It was constructed of wooden runners measuring 6" x 6" that span the length of the container. A staggered sheet of ¾" plywood was nailed and glued to the bottom of the runners to keep them stable. And attached to the bottom of the runners and plywood sheet were triple 2" x 6" pressure treated wooden risers that raise the height 6" for forklift forks to get under. The decking of the pallet was sheathed with double overlapped layers of ¾" plywood, with a layer of waterproof tar paper between the plywood layers, nailed and glued to the top of the runners. The pallet deck was a solidly built structure that was designed to support the heavy weight of the water system components. A post and rack system was nailed and bolted to the pallet deck in order to secure the water system equipment to the deck so it remained stable during transportation. The post and rack system was designed to distribute the weight of the water system equipment as evenly as possible on the pallet deck.

The lift-off cover was made with double overlapping ¼" sheets of plywood, also with a layer of tar paper between them. The plywood layers were joined together using 1" x 4" strips of lathing along the top, middle, and bottom on all sides of the plywood exterior. The vertical interior corners were stabilized with a post system of two 2" x 4" studs that were nailed together before the plywood was nailed to them. The bottom of the corner post was designed to rest on the protruding ends of the pallet runners. A single 2" x 4" was nailed to the horizontal interior corners of the cover and attached to the 2" x 4" were four brackets. These brackets were symmetrically spaced with two on each side along the length of the cover top, 8 to 12 feet apart. The cover overlapped the sides of the pallet deck and was held in place with 4" lag bolts approximately every 3 feet. A crane or large piece of excavating equipment was used to lift the cover onto (or off of) the pallet. Rope, straps, or chains of sufficient strength were attached to the brackets along the top and to the crane hook, and the cover was lifted off the deck. As a safety precaution, all exterior faces of the cover were coated with an earth-tone colored fire-resistant paint.

Engineers at PWS considered the design and construction of the container to be optimal for protecting the water system equipment from damage and theft during transportation and at the job site. The completed containers typically cost from $5,000 to $6,000 apiece, depending on the length, and the expense came from Mike's budget.

This was considered a negligible cost compared with the value of the equipment they held, although the cost of the container was factored into the price of the equipment. In addition, due to rapidly increasing lumber prices, the cost of a new container was creeping up to $7,000, and Mike was concerned that the cost might hit $8,000.

The Container Supply Chain

The loaded containers were moved onto a flatbed truck using a pair of industrial forklifts. The forks had to be long enough to extend under the container by a minimum of 5 feet or more and placed at least 4 feet apart. The total weight of a loaded container was kept under 30,000 pounds, and the value of the equipment ranged from $500,000 to $750,000, depending upon the application. It is important to note that the containers were not fully waterproof and that they were covered with a tarp while in transit. The container was then used as an onsite storage receptacle for the equipment for about one month while the installation work and system testing were being done. When the work was completed, the empty container was moved out of the way to reduce congestion in the work area.

Containers of water system equipment were shipped globally, but so far containers have only been returned from locations in the United States (U.S.) and Canada. Containers sent overseas (including Europe and Asia) have not come back. Mike was not sure what happens to them and was not aware of any product take-back laws in Europe that would apply to the containers. Containers were returned by U.S. and Canadian customers on a flatbed truck. Weight was not an issue; size was the issue, and Mike estimated that the average return cost was $1,900. Forecasting the return time of the containers was difficult due to construction project delays, and the condition of the already-returned containers currently onsite was not known.

One of the obvious problems with storing the containers at the Pasadena facility was the large size of the container. In addition, the containers were not stackable due to the minimal load-bearing capabilities in the walls and roof of the cover. So far, an excess of available outside space at the facility has allowed Mike to avoid a storage cost for the empty containers. But, Mike had been told that this cost-free arrangement will eventually cease to be an option and that an annual holding rate from 10% to 20% will soon be applied to the containers and charged to PPD. In addition, the containers are exposed to humid weather conditions and sooner or later they will begin to deteriorate. Mike estimated that the life expectancy of a container was between 18 to 24 months if it could be reused to ship water systems or to make smaller containers for water system replacement filters. Another problem was the growth of mold in an unknown number of containers while they were at the job site. These containers needed to be cleaned if they were to be reused. Finally, Mike estimated that depending on the length it would cost him from $500 to $1,000 to dispose of a container in a landfill.

Mike's Options

Mike believed he needed two value recovery solutions to his container problem. The first solution had to address the containers already onsite. Mike was willing to give the onsite containers to a third party who paid for shipping and assumed all liability for the containers once they were loaded onto the truck. The second solution had to focus on future returns. Preferably this solution would have the containers returned directly to a third-party buyer instead of to PWS and then to a buyer. In both cases, Mike would have to follow up on the exchange to make sure the receiving party did what they said they would do with the containers.

The construction of the containers placed constraints on how they could be reused. Though there was a lot of value in the timbers and plywood, it would take a significant amount of manual labor to break apart the container. The rack and post system anchored onto the pallet deck to secure the equipment was a major hindrance to both usage for other products and container disassembly.

Currently, Mike is able to use some of the containers (about 5%) to ship other products made at the plant, usually replacement filters. This entails cutting the container in sections and putting new ends on. Mike can make four small boxes from a 36- to 40-foot container. The cost to make the four small boxes is $1,500, and the cost to purchase one brand new small box is $1,000. So far, none of the small boxes have been returned. Mike contacted other manufacturers in the Pasadena-Houston metro area, but he was unable to find any buyers for the containers. Potential customers were leery of the container condition and the actual amount of work required to disassemble and retrofit the container because they did not know the container construction specifications.

Filling the containers with parts from PWS's suppliers for shipment to Pasadena was not an option because of the rack and post system bolted to the pallet deck. The cost to modify the container for supplier usage greatly exceeded the cost to purchase new containers. In addition, most parts were supplied in small quantities; therefore, filling the container was unlikely.

Selling the containers to a third-party company that made wood stove pellets was a possibility. However, it would have to be a small company that did not have access to waste from a paper mill. And, the company would have to strip off the paint and remove the tar paper between the sheets of plywood. Both of these were labor-intensive activities, and disposal of the stripped paint was an environmental issue because it could leach into the water supply if it was dumped in a landfill. The glue holding the plywood together might also be an environmental issue. Mike and his employees did not have enough knowledge of the wood pellet industry to accurately assess this option, and acquiring such knowledge would be time consuming.

Donations of the container to a nonprofit would be done at a very high corporate level, and the monetary effort was critical. The biggest issue with a charitable donation is the risk of exposure to mold that cannot be detected unless the container is taken apart. This potential problem also renders the containers unfit for human habitat. In addition, Mike would not see the charitable return on his income statement—it would be taken at the corporate level. Mike knew from prior experience with Pasadena Oil Equipment that used equipment was sold as is and with no liability or warranty. The sale also included a lengthy legal document. Mike believed that a donation to a charity would require an even longer legal document and this might extend the time it took to recover value from the existing containers. Also, Mike was not sure if he would be held responsible for the loading and shipping cost or if corporate would cover those expenses.

Reuse of the containers for the shipment of water systems was another option, but Mike was not sure how to repair or refurbish returned containers. A problem with reuse was how to grade the returned container and then determine the amount of material and labor needed to bring the container to a "good-as-new" specification. In order to provide a basis of discussion at the next meeting of his Product Packaging and Distribution (PPD) team, Mike developed a rough outline for condition and repair cost (Exhibit 2.1). He assumed a container would be ready for reuse within one month of its return. Mike was not aware of any existing cost-benefit or cost-avoidance models he could use to help him make a reuse decision; hence, he would have to develop one with his PPD team. Because PWS's P3 sustainability initiative was envisioned to be a partnership with suppliers and customers, Mike felt that customers who agreed to receive water system equipment in a used container should pay a reduced container price for their order. Given the extremely high value of the equipment, container reuse had a high level of risk associated with it.

Mike also wondered if they really needed to make 12 different sizes of containers and if a multi-use design would reduce the number of container models. This would require the pallet base to be made of high-strength steel while the pallet deck would remain wood. A problem with using steel or aluminum for the cover was that too much internal heat would be retained in the container, which could damage the equipment. The only way to reduce the internal temperature was to increase the number and size of the air vents, which would allow more moisture and possibly rain to enter the container. The current container had adequate ventilation, but it did not allow water inside. In addition, the use of steel would add to the shipping weight; currently, the empty containers weighed between 4,000 to 6,000 pounds. It was clear to Mike that the construction of the container had multiple constraints and redesign should be considered. However, redesign issues were currently not as important as recovering value from the existing containers.

Exhibit

Exhibit 2.1 Returned Container Condition and Repair Cost

Condition	Cost
No Repairs	$200 to inspect the condition and fumigate for mold
°Minor Repairs	< $1,000
°Medium Repairs	$1,000 to < $3,000
°Major Repairs	$3,000 to < $6,000
Scrap	≥ $6,000
°Convert to four crates	$1,500

* Includes inspection, fumigation, and material disposal costs from the rework.

3

Developing a Business Model to Improve the Energy Sustainability of Existing Buildings[1]

Barış Tan, Koç University

Emre N. Otay, Boğaziçi University

Yahya Yavuz, Koç University

Emre Çalıbel, Boğaziçi University

Introduction

Energy usage in buildings is responsible for approximately 33% of the total of final energy consumption in the world. 55% of the energy usage in buildings overall is at residential buildings and 45% of it is at nonresidential buildings, including commercial and public buildings. Heating, water heating, lights, and cooling are the primary sources of energy usage in buildings.

Natural gas and oil that are usually used for heating and cooling as well as electricity used for lights, cooling, refrigeration, and electronics play an important role in carbon dioxide (CO_2) emissions. Because emissions from burning fossil fuels are the primary cause of the rapid growth in atmospheric CO_2, it is possible to improve energy sustainability by increasing the energy efficiency in existing buildings.

Energy efficiency of buildings can be improved by implementing various energy efficiency measures. For example, previous studies on residential buildings show that it is possible to improve energy efficiency by insulating roofs (11% improvement), insulating external walls (30%), insulating cellar ceilings (6%), installing solar heating panels (7%), using energy efficient glazing (11%), or using a modern heating system (12%). Energy efficiency measures such as setting the domestic hot water system optimally or replacing electrical fixtures are also effective ways to improve energy efficiency of the buildings.

At the national level, improving energy efficiency of buildings can also be considered as a way of contributing to the future energy needs. Energy demand increases with the growth rate of countries. For fast-growing countries, there is a substantial demand for energy investments. Energy that can be saved in existing buildings can decrease the need for future energy investments.

Due to investment costs for installing and/or replacing technologies with more efficient ones and due to lack of expertise to identify the energy efficiency measures and implementing them effectively, energy efficiency projects are not implemented widely at existing buildings.

Implementing an energy efficiency measure at an existing building decreases energy consumption. Decreasing energy consumption reduces both future CO_2 emissions and also future energy expenditures. Savings in future energy expenditures can also be used to finance the initial investment. If an intermediary such as a service provider assumes the role of making the initial investment for customers with an agreement to share a fraction of future energy cost savings for a predetermined period, offering energy efficient technologies as a service can be a win-win-win arrangement for a service provider, its client, and also for the environment. The firm that offers the service can realize substantial financial returns. The customer pays a fraction of its energy bill with this agreement. Furthermore, realized energy savings will decrease CO_2 emissions and also ease the burden on future energy investments.

Improving Energy Efficiency at Boğaziçi Sarıtepe Campus

Sarıtepe Campus is a remote campus of Boğaziçi University located in Kilyos at the northern part of Istanbul on the Black Sea coast. The campus consists of seven buildings with a total flat area of 25,040 m^2 and 700 students.

The average natural gas and electricity consumption and the current CO_2 emissions for this campus are given in Exhibit 3.1. Energy use intensity is 130kWh/m^2 and the carbon emission per unit area is 45kg CO_2/m^2. With these values, its current energy level is evaluated as level D.

As a part of the Sustainable and Green Campus Initiative for Boğaziçi University, Emre N. Otay, professor of civil engineering and the coordinator of the Boğaziçi Sarıtepe Campus, decided to start a project to improve the energy efficiency of the campus.

The energy efficiency of the campus can be increased by improving the natural gas heating system, retrofitting the exterior walls, adding insulation to the roofs, improving the lighting system, and creating sunrooms on roofs and balconies to improve energy efficiency in a passive way by using architectural design. However, each alternative requires an investment and leads to a particular level of energy saving, which, therefore, leads to a cost savings and also a reduction in CO_2 emissions.

Emre N. Otay explains,

We would like to use the budget that will be allocated to this project in the best way possible to achieve the maximum benefit. Therefore, selecting the energy efficiency measures within our budget in the best way possible is a crucial task for us. However, what is more challenging is determining the energy savings that will be achieved by implementing a given energy efficiency measure.

Energy Efficiency Measures Identified for Boğaziçi Sarıtepe Campus

The challenge raised by Professor Otay led to a wider study beyond the scope of the project. University campuses are small-scale, self-contained living environments where the residents spend their complete daily life cycle within the buildings at the site. Furthermore, analyzing a campus complex instead of a single stand-alone building allows the inclusion of existing buildings of different sizes, layouts, functions, and levels of energy efficiency; the effect of buildings on each other; and the impact of common areas and common mechanical systems on the energy efficiency of the whole complex in the analysis. As a result, it is possible to use the insights from the analysis of a university campus to design systems for larger living environments.

During the same period, Emre Çamlıbel, CEO of Soyak Holding who received a master's degree from Massachusetts Institute of Technology (MIT) before starting his business career, went back to the graduate school to pursue a doctorate degree in civil engineering. Soyak Holding is a holding company that specializes in the construction and energy sectors. As the top executive of a group that builds LEED-certified green buildings as well as power plants, he was the ideal student to meet the challenge raised by Professor Otay. This project within the context of the wider research problem became the doctorate dissertation of Emre Çamlıbel.

By using his architectural and engineering expertise, Emre Çamlıbel identified investment costs and the expected energy savings from each energy-saving technology through detailed technical and mechanical modeling and measurements. A specific software (IZODER TS 825) was used for calculating the shell U-values of the buildings and their heating energy demand. LEED and Energy Star software was used to assess the energy consumption and CO_2 emission levels with the implementation of a given energy efficiency measure.

Exhibit 3.2 lists energy efficiency measures (EEM) that can be implemented in seven different buildings on the campus. Exhibit 3.3 shows the investment cost (in $) and the energy saving (in kWh) for each energy efficiency measure that is identified for the campus. Exhibit 3.3 also indicates the type of energy savings (heating or electric energy) and the source of energy affected by using each technology. Because natural gas is used for heating, a conversion factor of 0.234kg/kWh can be used to determine the amount of CO_2 emission due to natural gas consumption, and a conversion factor of 0.617kg/kWh can be used for CO_2 emission due to electric energy consumption. The financial savings is based on whether the proposed energy efficiency measure improves heating energy obtained from natural gas or electric energy consumption. The electricity price at the time of the analysis was 0.134$/kWh, and the price of natural gas was 0.037$/kWh.

Because natural gas is used for both air heating (EEM #19 to #42 in Exhibits 3.2 and 3.3) and also for water heating (EEM #1 to #14) and the price for a cubic meter (m^3) of natural gas is the same, the efficiency of the boiler used to heat water yields a higher cost per kWh of heating energy used for water heating. The efficiency of the boiler used at the campus that is measured to be 90.17% compared with air heating can be used to determine this cost.

Among the energy efficiency measures identified, the measures related to envelope insulation retrofit with 4cm, 5cm, and 6cm options (EEM #19 to #36) for a given building are mutually exclusive. In other words, only one of these options can be implemented for a given building.

Selecting the Energy Efficiency Measures Optimally for a Given Budget

Once the investment amounts, energy, cost, and CO_2 savings are identified for each energy efficiency measure, the question that needs to be answered is the following: *If you have a given budget to invest in energy efficiency measures, in which technologies should you invest in order to maximize reductions in energy usage, CO_2 emissions, or cost savings?*

The relative importance of these measures is different for different organizations; using different objective functions yields a different set of recommended energy efficiency measures.

Emre N. Otay says,

> For us, using the allocated budget to minimize the CO_2 emissions is the primary objective according to our Sustainable and Green Campus Initiative. Future energy cost savings are good to have but not the main objective. Saving energy is also important, but if I have a choice I prefer the ones that reduce CO_2 emissions.

Emre Çamlıbel disagrees,

> Three objectives, maximizing reductions in CO_2 emissions, maximizing cost savings, and maximizing energy savings, are interrelated: The source of emissions savings is reduction in energy consumption. If we maximize cost savings, we will also

improve reductions in CO_2 emissions and energy usage. The difference between maximizing cost savings and CO_2 emission savings is not expected to be significant for the amount of CO_2 emission that will be saved. Furthermore we can use the additional cost savings for other sustainability and green campus initiatives. So our objective should be maximizing cost savings.

A mathematical program can be formulated to select the energy efficiency measures within a given budget to optimize a given objective function subject to constraints that may limit the selection of some of the energy efficiency measures jointly. Then, the energy, cost, and CO_2 savings obtained by using different objective functions can be compared to make a recommendation to Emre N. Otay regarding which energy efficiency measures should be selected.

Selecting Energy Efficiency Measures in Multiple Time Periods

Once the data that shows the required investments and potential savings for each energy efficiency measure is obtained and an optimization model is developed to select the energy efficiency measures to optimize a given objective function for a given year, the same model can be extended to analyze the problem in a multi-period setting.

In a multi-period setting, after the initial investment, future savings in energy expenditures can be used to invest in other energy efficiency measures or to accumulate cash for investments in later periods. By following this method, substantial savings can be achieved by using a relatively low level of initial investment. In this case, the decision variables include both selection of the technology and also timing of the investment in each selected technology. The money saved compared with the previous years' expenditures and not used at the end of a given year can be invested in short-term financial markets.

The proposed multi-period model takes the approach that the future financial savings resulting from improving energy efficiency can be directed as a fund for investing only in other energy efficiency measures or in the short-term financial market.

Emre Otay asks,

For us, we know how much we pay for energy each year. If we decrease the energy cost next year, it is not an additional financial fund we can invest. In this case, how can we use this approach?

The answer may be using a long-term contract with an energy provider to set the energy price. In this case, savings compared with what is budgeted according to the long-term contract can be directed to other investments.

Emre Çamlıbel adds,

The business model that is used by Energy Service Companies (ESCOs) depends on offering investments in energy efficiency measures as a service. If such a business model can be developed in Turkey, it can be a lucrative business while improving energy efficiency of existing buildings.

Offering Energy Efficiency Measures as a Service: ESCO Business Model

In the ESCO business model, a service provider (ESCO) offers to make all of the necessary energy saving technology investments for a client in exchange for receiving a fraction of the savings in energy expenditures for a predetermined time period. In other words, the total energy savings from investing in energy efficiency measures in each period are shared between the client that receives a part of the total savings and the service provider that receives the remaining part.

For the success of this business model, the right set of energy efficiency measures must be selected given the budgetary constraints and the objectives regarding CO_2 emissions and financial returns.

The parameters of the contract—that is, the fraction of the savings that will be kept by the service provider and the contract period—must be set correctly. For a given set of parameters, the service provider should also determine the right amount of investment in energy efficiency measures to achieve the energy, cost, and CO_2 savings for the client while obtaining a desired level of return for its investment.

The mathematical programming approach used for the single period can be used to select the best set of technologies.

Investing in Energy Efficiency Measures as an Alternative Source of Energy

As a fast-growing country, Turkey is expected to grow around 5% annually during the next five years. Accordingly, the energy demand is increasing rapidly. Turkish gross domestic electricity consumption is expected to grow 7.5% annually, and the natural gas demand is expected to grow around 3% annually during the 2012–2020 period. In this setting, the number of investments in energy production is accelerating.

Emre Çamlıbel comments,

> The investment cost of a hydroelectric power plant recently built in Turkey that will produce 144 million kWh of energy in a year was $96 million. If the same investment was directed to investing in improving energy efficiency of existing buildings, similar to Sarıtepe campus, the total energy savings from all the projects that use the investment of a single hydroelectric plant will bring more energy compared to building

hydroelectric plants. So for an energy investor's point of view, this can be an even more attractive source of energy.

There are close to 14 million residential buildings in Turkey, and 92% of them do not comply with the energy efficiency standards. Every year, approximately 500,000 new buildings are built in the country. If the useful life of a building is 50 years, it will take 30 years for half of the total number of buildings to be new and comply with the standards. So, it may be possible to implement investments in energy efficiency measures at a large scale by using a well-designed business model and create a win-win-win situation for the service providers, building owners, and for the environment.

Study Questions

1. Determine the annual savings in energy costs and CO_2 emissions for each energy efficiency measure.
2. Formulate a mathematical program to select energy efficiency measures optimally for a given budget subject to the constraints that limit selection of certain energy efficiency measures.
3. Compare the optimal selection of energy efficiency measures using maximizing cost savings, maximizing energy savings, and maximizing energy savings as the main objective function. What are the differences and similarities between the sets of optimal measures? Which energy efficiency measures would you recommend using for this project?
4. Suppose that you take the 96 million USD required to build a hydroelectric plant that will produce 144 million kWh/year and instead invest those funds in different energy efficiency projects for existing buildings. How much energy can you save per year? (Assume that you invest 100,000 USD in each project and

the characteristics of each project are identical to the Sarıtepe Campus.)

5. Formulate a mathematical program to select the optimal set of energy efficiency measures in a multi-period setting. All the money not used in a given year can be invested in financial markets at an interest rate of 2% per year. The service provider expects an annual return that is more than the interest rate. Use the first ten energy efficiency measures and consider a seven-year planning horizon. Assume that the energy cost savings stay the same during the planning horizon.

6. Consider a business plan to offer investing in energy efficiency measures as a service. The service provider makes the necessary initial investment and collects an agreed-upon percentage of cost savings from the customer. The service provider requires a minimum of 2% annual return from this investment.

    a. Select the energy efficiency measures to be invested and the timing of these investments in a seven-year period to maximize the financial savings, energy savings, and CO_2 savings if the company wants to invest 100,000 USD and the agreed-upon share of savings is 45%.

    b. What is the optimal level of investment to maximize financial returns for the service provider?

    c. Which factors affect the feasibility of this business plan? Discuss these factors and their implications.

Endnote

1. Also available in The Case Centre Collection (Reference No. 614-074-1).

Exhibits

Exhibit 3.1 Energy Consumption and CO_2 Emissions for Boğaziçi Sarıtepe Campus

	Electricity	Natural Gas	Total
Total consumption (kWh)	979,480	2,276,839	3,256,319
Total CO_2 emission (kg)	604,339	532,780	1,137,119

Exhibit 3.2 Energy Efficiency Measures That Can Be Implemented in Different Buildings

EEMS	Dorm I N-Block	Dorm I S-Block	Dorm I Fac. Apts.	Prep Sch Bldg A	Prep Sch Bldg B	Hotel	Dorm II
Optimization of domestic hot water system	D1 (1)	D1 (1)	D1 (1)	–	–	H1 (2)	I1 (3)
Heating system piping insulation	A2 (4)	B2 (5)	C2 (6)	E2 (7)	F2 (8)	–	I2 (9)
Renovation of boiler	D3 (10)	D3 (10)	D3 (10)	–	–	–	–
Installation of thermostatic valves	A4 (11)	B4 (12)	C4 (13)	–	–	–	I4 (14)
Change of lightbulbs' ballasts	D5 (15)	D5 (15)	D5 (15)	G5 (16)	G5 (16)	H5 (17)	I5 (18)
Envelope insulation environments—6cm	A6 (19)	B6 (20)	C6 (21)	E6 (22)	F6 (23)	–	I6 (24)
Envelope insulation environments—5cm	A7 (25)	B7 (26)	C7 (27)	E7 (28)	F7 (29)	–	I7 (30)
Envelope insulation environments—4cm	A8 (31)	B8 (32)	C8 (33)	E8 (34)	F8 (35)	–	I8 (36)
Installation of variable-speed drive pumps	D9 (37)	D9 (37)	D9 (37)	G9 (38)	G9 (38)	–	I9 (39)
Trombewall application	A10 (40)	B10 (41)	–	–	–	–	–
Creating sunrooms on roofs and balconies	–	–	–	E10 (42)	–	–	–

Exhibit 3.3 Investment Costs and the Annual Energy Savings for the Energy Efficiency Measures

No.	Code	Energy Type	Source	Investment in $	Energy Saving in kWh/year
1	D1	Heating	Natural gas (water heating)	1,250	76,827
2	H1	Heating	Natural gas (water heating)	500	3,295
3	I1	Heating	Natural gas (water heating)	1,250	55,557
4	A2	Heating	Natural gas (water heating)	1,071	23,443
5	B2	Heating	Natural gas (water heating)	1,071	23,443
6	C2	Heating	Natural gas (water heating)	964	21,099
7	E2	Heating	Natural gas (water heating)	2,330	50,989
8	F2	Heating	Natural gas (water heating)	6,750	229,189
9	I2	Heating	Natural gas (water heating)	1,969	43,077
10	D3	Heating	Natural gas (water heating)	41,250	250,005
11	A4	Heating	Natural gas (water heating)	5,850	61,050
12	B4	Heating	Natural gas (water heating)	5,850	60,764
13	C4	Heating	Natural gas (water heating)	1,300	17,402
14	I4	Heating	Natural gas (water heating)	9,100	63,969
15	D5	Electricity	Electricity	10,938	24,599
16	G5	Electricity	Electricity	12,500	37,426
17	H5	Electricity	Electricity	1,563	1,972
18	I5	Electricity	Electricity	6,250	17,200
19	A6	Heating	Natural gas (air heating)	34,402	95,692

No.	Code	Energy Type	Source	Investment in $	Energy Saving in kWh/year
20	B6	Heating	Natural gas (air heating)	34,402	95,611
21	C6	Heating	Natural gas (air heating)	17,084	27,329
22	E6	Heating	Natural gas (air heating)	33,448	156,501
23	F6	Heating	Natural gas (air heating)	22,182	69,278
24	I6	Heating	Natural gas (air heating)	38,998	92,882
25	A7	Heating	Natural gas (air heating)	30,850	81,789
26	B7	Heating	Natural gas (air heating)	30,850	81,210
27	C7	Heating	Natural gas (air heating)	15,320	23,286
28	E7	Heating	Natural gas (air heating)	29,995	152,049
29	F7	Heating	Natural gas (air heating)	19,892	59,272
30	I7	Heating	Natural gas (air heating)	34,972	79,314
31	A8	Heating	Natural gas (air heating)	27,055	61,993
32	B8	Heating	Natural gas (air heating)	27,055	61,948
33	C8	Heating	Natural gas (air heating)	13,436	17,706
34	E8	Heating	Natural gas (air heating)	26,305	145,914
35	F8	Heating	Natural gas (air heating)	17,445	45,575
36	I8	Heating	Natural gas (air heating)	30,670	60,237
37	D9	Electricity	Natural gas (air heating)	3,750	4,455
38	G9	Electricity	Natural gas (air heating)	6,250	11,880
39	I9	Electricity	Natural gas (air heating)	3,750	5,940

No.	Code	Energy Type	Source	Investment in $	Energy Saving in kWh/year
40	A10	Heating	Natural gas (air heating)	65,655	31,912
41	B10	Heating	Natural gas (air heating)	65,655	31,890
42	E10	Heating	Natural gas (air heating)	119,824	65,837

4

Contracting for LTL Services at Hankey Industries

Matthew J. Drake, Duquesne University

Michael Sanger arrived at his small, windowless office stuck into a corner of the Hankey Industries manufacturing plant in Catasaqua, Pennsylvania,[1] just after 8:00 a.m. on Monday, June 16. He opened his email Inbox and uttered a small sigh as he saw emails from three sales representatives for LTL carriers containing large attachments. The sigh signified the amount of work that he knew these emails had in store for him over the next few weeks.

Michael was the logistics manager for Hankey Industries, a small manufacturer of cleaning and specialty chemicals with annual sales of $8.5 million. Hankey Industries was founded in Catasaqua, Pennsylvania, in 1949 by Joshua Hankey, a veteran of the U.S. Navy in the Pacific theater of World War II. Joshua's primary goal was to help his customers solve their difficult industrial cleaning and specialty chemical problems, a commitment to customer service that has continued to this day. The current manufacturing plant is located adjacent to the original site in a larger building that opened in 1978.

Hankey Industries' Logistics Operations

Hankey Industries offers a variety of industrial cleaning and specialty chemical products, mainly through distributors located in the eastern United States, although it does provide some cleaners directly to large railroads, which use the products to clean their locomotives. As a result, Hankey's logistics operations are relatively simple to map. Most shipments travel from the plant in Catasaqua to a small number of customer locations, which represent the distributors' warehouses or the railroads' maintenance centers. A list of Hankey's major customers' locations is provided in Exhibit 4.1.

Because its products are relatively homogeneous with respect to their physical characteristics (such as density, packaging, susceptibility to damage, etc.), Hankey's Logistics Department has developed a standard pallet size with a total weight of 400 pounds. Most customers, especially the distributors, order full pallet quantities of each product at regular intervals (such as every week or every month). Hankey provides pricing incentives to encourage its customers to order mixed pallets of products at the standard size of 400 pounds per pallet whenever the customers do not require a full pallet of a single product. Sometimes, however, a customer finds it necessary to place an order for a quantity smaller than a full pallet. This could be because the customer does not have the storage space to hold an additional full pallet of product and can only accept a partial pallet or because usage of the product is too slow relative to the size of a full pallet. In line with Hankey's stated commitment to customer service, these requests are handled on a case-by-case basis to meet the customer's needs.

Hankey Industries' operations are small enough that its entire Logistics Department consists of Michael (the department manager) and two traffic associates who handle the day-to-day operations, such as tendering shipments to carriers and tracing shipments that were not delivered on time, among their other tasks. There are no dedicated logistics analysts in the department; as a result, Michael had to perform all of the analysis himself.

Contracting for LTL Services

Hankey Industries' existing contract for LTL transportation with American Freightways was expiring as of June 30. A few weeks ago Michael contacted his sales representative at American Freightways as well as the representatives at three other LTL carriers with a request for a new LTL pricing contract that would be effective for the next 18 months (from July 1 of the current year to December 31 of the following year).[2] He prepared a representative list of Hankey's LTL shipments from April 1 to April 30 of the current year (provided in Exhibit 4.2) to accompany this request.[3] As Hankey's business is relatively stable throughout the year, this one-month data set would give each carrier an accurate forecast of the volume of shipments that it would see from Hankey Industries if it won the contract. The homogeneity of Hankey's products with respect to density, packaging, and susceptibility to damage allows for all of its products to be rated at LTL freight class 55 according to the National Motor Freight Classification. The American Freightways representative sent his quote to Michael within a few days, but the other carriers' representatives took until a few days ago to email their quotes.

Each of the carriers submitted proposals following the guidelines Michael established in his initial request. Based on the sample month of shipments provided in Exhibit 4.2, the carriers' representatives developed several types of rate quotes for shipments moving from Catasaqua to each of the 15 customer locations in Exhibit 4.1:

1. A set of quotes that apply to all 15 customer locations if the carrier is selected as Hankey's sole source of contracted LTL transportation services
2. A set of quotes that could be selected individually for any of the 15 customer locations

3. Three sets of combination quotes containing rates that can only be selected together (e.g., special pricing for the two Ohio locations that applies when both locations are contracted in tandem)

Each rate quote prescribes a base rate per hundredweight (cwt) for a range of shipping volumes at LTL freight class 55, a corresponding discount percentage, and a fuel surcharge percentage. All of Hankey's shipments are large enough and must travel a long enough distance that none of the carriers' minimum charges would apply to any of these regular shipments. Exhibits 4.3 through 4.6 summarize the base rates and the discounts submitted by each of the four potential carriers.

LTL Rate Quotes

The American Freightways quote is provided in Exhibit 4.3. In addition to the base rates and discount percentages listed in the table, American Freightways charges a fuel surcharge of 29% applied to the discounted freight charges. The quote also contains the following location combinations that only apply if they are contracted together:

- **Combination 1**—Gahanna, OH (63% discount) and Calcutta, OH (63% discount)
- **Combination 2**—Charlotte, NC (65% discount) and Gaffney, SC (67% discount)
- **Combination 3**—Worcester, MA (65% discount), South Plainfield, NJ (66% discount), and Cherry Hill, NJ (65% discount)

Exhibits 4.4 through 4.6 detail the quotes submitted by three carriers with which Hankey Industries does not currently do business—Consolidated Freightways, Parker Motor Freight, and A-P-A Transport. The following list provides the fuel surcharge percentages and combination quotes that each of the carriers has submitted.

- Consolidated Freightways (Exhibit 4.4)
  - Fuel surcharge: 28%
  - **Combination 1**—Altoona, PA (63% discount) and Sabillasville, MD (66% discount)
  - **Combination 2**—Wytheville, VA (68% discount), Nitro, WV (67% discount), and Huntington, WV (68% discount)
  - **Combination 3**—Waycross, GA (65% discount) and Cleveland, TN (69% discount)
- Parker Motor Freight (Exhibit 4.5)
  - Fuel surcharge: 30%
  - **Combination 1**—South Plainfield, NJ (67% discount) and Cherry Hill, NJ (64% discount)
  - **Combination 2**—Worcester, MA (64% discount) and Rotterdam, NY (65% discount)
  - **Combination 3**—Cleveland, TN (67% discount), Gahanna, OH (62% discount), and Calcutta, OH (61% discount)
- A-P-A Transport (Exhibit 4.6)
  - Fuel surcharge: 29%
  - **Combination 1**—Sabillasville, MD (66% discount) and Wytheville, VA (69% discount)
  - **Combination 2**—Nitro, WV (65% discount), Huntington, WV (68% discount), and Calcutta, OH (65% discount)
  - **Combination 3**—Cherry Hill, NJ (65% discount), Charlotte, NC (67% discount), and Gaffney, SC (70% discount)

All of these contract proposals follow the LTL pricing convention of charging for deficit weight when a shipment's weight approaches a new rate level (known as a *rate break*). The carrier always charges the lower of the shipment's actual weight priced at the corresponding rate per cwt and the weight at the rate break priced at the rate break rate. For example, suppose that shipments weighing less than

1,000 pounds are priced at $100 per cwt and shipments weighing 1,000 pounds or more are priced at $90 per cwt. A shipment weighing 950 pounds would be rated as if it weighed 1,000 pounds because the base rate at the actual weight (9.5 × $100 = $950) is larger than the base rate at the rate break (10 × $90 = $900). The base rate for this shipment would include a "deficit weight" of 50 pounds to make the total rated weight of the shipment equal to 1,000 pounds. Any analysis of an LTL rate proposal must begin with the determination of the break point (in weight) where each rate break applies.

Selecting Contracts for LTL Services

After receiving the rate quotes from the four LTL carriers, Michael planned the steps of the analysis that would consume the majority of his next two weeks in the office. Once he determined the weight at which each rate break applied, he must then rate every shipment in his sample data set under each contract option offered by each carrier. Finally, he must optimize the selection of contracts to ensure that every location is covered by one carrier and that the total transportation cost is minimized.

With the plan in place, Michael began his analysis. Time was of the essence if he wanted to have the new contracts in place by the end of June. He wondered how much the company could save each month by working with a combination of carriers instead of contracting only with one. The potential savings would have to be large enough to justify the additional hassle of having his traffic associates contact multiple carriers each day to tender shipments. There would also be an increased risk that the associate would tender the shipment to the wrong carrier for a particular customer. Michael was eager to see how the analysis came out.

Endnotes

1. The zip code for Catasaqua, Pennsylvania, is 18032.
2. The names of all four LTL carriers used in this case are an homage to real LTL carriers that are no longer in operation. The carriers were either purchased by other carriers or went out of business in the decade of the 2000s.
3. As he prepared this data set, he eliminated any shipments that would be considered "irregular" because they were unusually large or small or because they were destined for locations to which Hankey Industries rarely shipped. The shipment data set in Exhibit 4.2 only contains regular shipments to common locations, which serve as the basis for any long-term contract.

Exhibits

Exhibit 4.1 List of Hankey Industries' Major Customer Locations

Customer Type	City	State	Zip Code
Distributor warehouse	Worcester	MA	01608
Distributor warehouse	South Plainfield	NJ	07080
Distributor warehouse	Cherry Hill	NJ	08034
Distributor warehouse	Rotterdam	NY	12306
Railroad maintenance	Altoona	PA	16601
Distributor warehouse	Sabillasville	MD	21780
Distributor warehouse	Wytheville	VA	24382
Distributor warehouse	Nitro	WV	25143
Railroad maintenance	Huntington	WV	25701
Railroad maintenance	Charlotte	NC	28202
Distributor warehouse	Gaffney	SC	29340
Railroad maintenance	Waycross	GA	31503
Distributor warehouse	Cleveland	TN	37312
Distributor warehouse	Gahanna	OH	43230
Distributor warehouse	Calcutta	OH	43920

Exhibit 4.2 Hankey Industries' LTL Shipments in April

Date	Destination Zip	Number of Pieces	Total Weight
1-Apr	12306	3	1,050
1-Apr	43920	1	400
2-Apr	25143	4	1,600
2-Apr	29340	6	2,400
2-Apr	43230	7	2,800
3-Apr	01608	5	1,845
3-Apr	07080	3	1,200
4-Apr	24382	4	1,600
4-Apr	37312	2	800
7-Apr	08034	4	1,600
7-Apr	25701	5	1,910
8-Apr	12306	2	800
8-Apr	43920	5	1,885
9-Apr	25143	3	975
9-Apr	29340	4	1,600
9-Apr	43230	11	4,400
10-Apr	07080	4	1,600
10-Apr	28202	6	2,400
11-Apr	24382	9	3,480
11-Apr	37312	1	400
14-Apr	08034	4	1,600
15-Apr	07080	2	625
15-Apr	12306	4	1,600
16-Apr	21780	3	1,095
16-Apr	29340	7	2,800
16-Apr	43920	3	1,200
17-Apr	01608	1	400
17-Apr	25143	2	800
17-Apr	43230	8	3,200
18-Apr	24382	3	1,120
18-Apr	37312	2	800
21-Apr	08034	7	2,800
21-Apr	43920	3	1,200
22-Apr	12306	9	3,600

Chapter 4 • Contracting for LTL Services at Hankey Industries

Date	Destination Zip	Number of Pieces	Total Weight
22-Apr	25143	5	2,000
23-Apr	07080	3	1,200
23-Apr	29340	6	2,400
23-Apr	43230	5	2,000
24-Apr	01608	3	1,200
24-Apr	37312	4	1,505
25-Apr	24382	5	2,000
25-Apr	43920	6	2,310
28-Apr	07080	2	800
28-Apr	08034	3	1,200
29-Apr	25143	3	1,200
29-Apr	31503	8	3,200
29-Apr	43230	9	3,465
30-Apr	12306	6	2,400
30-Apr	16601	2	800
30-Apr	21780	3	990
30-Apr	29340	4	1,445

Exhibit 4.3 American Freightways' Rate Quote

Customer Zip Code	Base Rate (in $/cwt) for Shipments with Weight W					
	W < 500 lbs	500 <= W < 1,000 lbs	1,000 <= W < 2,000 lbs	2,000 <= W < 5,000 lbs	5,000 <= W < 10,000 lbs	
01608	129.65	95.54	72.76	60.42	48.75	
07080	128.28	98.87	74.14	60.01	45.63	
08034	109.35	84.32	62.75	50.01	37.03	
12306	118.83	91.76	69.45	56.77	44.88	
16601	111.82	86.40	65.54	54.04	42.77	
21780	103.68	79.95	60.45	49.18	38.23	
24382	134.00	104.44	80.20	67.79	56.53	
25143	156.38	122.07	93.99	79.86	67.31	
25701	157.47	122.80	94.69	80.75	68.10	
28202	110.49	101.64	79.57	65.18	50.83	
29340	119.38	109.87	85.94	70.41	54.93	
31503	140.60	129.26	103.99	88.59	68.85	
37312	138.45	127.38	102.39	87.24	67.81	
43230	124.16	105.49	88.16	78.19	62.04	
43920	112.60	95.74	79.99	70.94	56.35	

CHAPTER 4 • CONTRACTING FOR LTL SERVICES AT HANKEY INDUSTRIES

Customer Zip Code	Discount if Sole Source	Discount for Individual Location	Combination 1 Discounts	Combination 2 Discounts	Combination 3 Discounts
01608	68%	61%			65%
07080	69%	65%			67%
08034	68%	62%			66%
12306	63%	59%			
16601	66%	60%			
21780	64%	61%			
24382	64%	60%			
25143	65%	62%			
25701	64%	59%			
28202	67%	61%		65%	
29340	68%	64%		66%	
31503	65%	62%			
37312	65%	61%			
43230	66%	58%	65%		
43920	68%	60%	67%		

Exhibit 4.4 Consolidated Freightways' Rate Quote

Customer Zip Code	Base Rate (in $/cwt) for Shipments with Weight W					
	W < 500 lbs	500 <= W < 1,000 lbs	1,000 <= W < 2,000 lbs	2,000 <= W < 5,000 lbs	5,000 <= W < 10,000 lbs	
01608	130.95	96.50	73.49	59.21	47.78	
07080	130.85	100.85	75.62	59.41	45.17	
08034	112.63	86.85	63.38	50.51	37.40	
12306	117.64	90.84	68.76	54.50	43.08	
16601	110.70	85.54	66.20	54.58	43.20	
21780	105.75	81.55	61.66	49.67	39.00	
24382	131.32	102.35	78.60	66.43	57.66	
25143	161.07	125.73	93.99	80.66	66.64	
25701	155.90	121.57	92.80	79.14	66.79	
28202	111.04	102.15	79.17	64.85	50.58	
29340	120.57	110.97	88.52	72.52	56.58	
31503	134.98	124.09	100.87	85.93	66.78	
37312	139.83	128.65	104.44	88.29	68.83	
43230	125.90	106.97	89.13	77.41	61.42	
43920	112.71	96.22	80.63	69.52	53.53	

Chapter 4 • Contracting for LTL Services at Hankey Industries

Customer Zip Code	Discount if Sole Source	Discount for Individual Location	Combination 1 Discounts	Combination 2 Discounts	Combination 3 Discounts
01608	66%	60%			
07080	65%	62%			
08034	67%	63%			
12306	64%	58%			
16601	63%	59%	62%		
21780	70%	64%	69%		
24382	70%	65%		68%	
25143	70%	63%		69%	
25701	72%	65%		70%	
28202	65%	61%			
29340	66%	64%			
31503	69%	60%			68%
37312	71%	65%			69%
43230	60%	56%			
43920	63%	59%			

Exhibit 4.5 Parker Motor Freight's Rate Quote

Base Rate (in $/cwt) for Shipments with Weight W

Customer Zip Code	W < 500 lbs	500 <= W < 1,000 lbs	1,000 <= W < 2,000 lbs	2,000 <= W < 5,000 lbs	5,000 <= W < 10,000 lbs
01608	128.35	94.58	72.03	61.63	49.73
07080	125.71	96.89	74.88	60.61	46.09
08034	112.63	86.85	61.50	49.01	36.29
12306	120.02	92.68	70.14	57.34	43.53
16601	106.23	82.08	62.26	54.58	43.20
21780	107.83	83.15	62.87	50.16	38.99
24382	132.66	103.40	76.99	65.08	54.27
25143	151.69	118.41	91.17	80.26	67.65
25701	160.62	125.26	96.58	81.80	68.99
28202	113.80	104.69	77.74	63.68	49.42
29340	114.60	105.48	81.64	66.89	52.18
31503	139.19	127.97	102.95	87.70	68.16
37312	142.60	131.20	104.44	88.98	69.17
43230	125.40	106.54	89.04	81.32	64.52
43920	109.22	92.87	77.59	68.81	55.79

Chapter 4 • Contracting for LTL Services at Hankey Industries

Customer Zip Code	Discount if Sole Source	Discount for Individual Location	Combination 1 Discounts	Combination 2 Discounts	Combination 3 Discounts
01608	65%	60%		64%	
07080	69%	63%	67%		
08034	68%	61%	66%		
12306	66%	60%		65%	
16601	64%	58%			
21780	70%	63%			
24382	69%	64%			
25143	65%	61%			
25701	68%	63%			
28202	65%	60%			
29340	67%	64%			
31503	64%	60%			
37312	70%	63%			67%
43230	64%	58%			62%
43920	63%	59%			61%

Exhibit 4.6 A-P-A Transport's Rate Quote

	Base Rate (in $/cwt) for Shipments with Weight W				
Customer Zip Code	W < 500 lbs	500 <= W < 1,000 lbs	1,000 <= W < 2,000 lbs	2,000 <= W < 5,000 lbs	5,000 <= W < 10,000 lbs
01608	133.54	98.41	74.94	61.02	49.24
07080	130.97	100.95	74.96	61.10	46.00
08034	104.87	80.86	60.18	48.51	36.29
12306	115.86	89.47	67.71	55.35	43.76
16601	114.62	87.35	66.26	55.07	43.75
21780	104.41	80.51	59.24	48.39	37.58
24382	135.34	106.01	81.16	68.54	57.60
25143	159.82	124.51	93.05	77.46	63.94
25701	152.75	120.22	92.51	79.94	66.26
28202	108.28	99.81	78.69	66.48	50.32
29340	121.05	111.30	87.57	70.55	55.70
31503	141.16	131.07	104.93	88.15	69.13
37312	137.07	127.13	101.67	85.67	67.13
43230	119.44	102.22	84.63	75.69	60.74
43920	113.61	96.70	80.31	72.29	57.48

Chapter 4 • Contracting for LTL Services at Hankey Industries

Customer Zip Code	Discount if Sole Source	Discount for Individual Location	Combination 1 Discounts	Combination 2 Discounts	Combination 3 Discounts
01608	63%	59%			
07080	66%	63%			
08034	67%	60%			66%
12306	63%	58%			
16601	62%	57%			
21780	69%	63%	68%		
24382	71%	65%	70%		
25143	66%	61%		65%	
25701	68%	64%		67%	
28202	70%	63%			
29340	73%	66%			69%
31503	63%	59%			72%
37312	65%	61%			
43230	61%	57%		65%	
43920	67%	60%			

5

Optimal Product Bundling at Point and Shoot Camera Shop

Matthew J. Drake, Duquesne University

Steve Bennett is the owner of the Point and Shoot Camera Shop, a small chain of camera stores located throughout Ohio, Pennsylvania, and Indiana. In 2010, Steve turned 68 years old and decided to retire from day-to-day operations. He transferred control of the company to Cheryl Peretin, the newly appointed president and CEO, and feels comfortable that the company he worked so hard to build from the ground up is in safe hands. Cheryl had worked for Steve for 17 years and had always proven herself to be a strong leader and an effective manager.

Company Background

Steve founded the original store location in Kent, Ohio, in 1979 to cater to the growing number of photography enthusiasts in the area, which was largely fueled by the successful photography programs at nearby Kent State University and the University of Akron. Steve was a former engineer who quit his job in the electrical industry to follow his passion for photography by opening the store. He felt a great deal of satisfaction when he was able to match a customer with the perfect camera for his or her needs and level of expertise.

In the subsequent years, Steve expanded the brand by opening new stores in cities with major colleges, such as Bloomington, Indiana; Columbus, Cincinnati, and Toledo, Ohio; and Pittsburgh and Erie, Pennsylvania. The chain currently operates 11 stores in these three states and generates a total of approximately $15 million in annual sales.

Point and Shoot's most popular camera style is an entry-level digital SLR model. This is a perfect model for new photography enthusiasts who are looking to graduate from the cheap, pocket-sized digital cameras or smartphone cameras to a more sophisticated offering. This model is also recommended by the photography departments at many of the nearby universities for incoming students to own when they enter the program. In fact, the chain has agreements with many of the departments to offer their students a discount on this specific model to encourage them to purchase the camera that they need from Point and Shoot.

Cheryl had been very successful in her first several years at the head of the company. She had overseen the opening of two new stores in Fort Wayne, Indiana, and State College, Pennsylvania, in 2014; most recently, she spearheaded the introduction of photo-printing kiosks in all of the stores to expand the services the stores offer to their customers. She has also maintained and even expanded in some locations the popular classes that teach novice customers how to use their new digital SLR cameras.

Product Bundling

Cheryl made it a point to check in with Steve weekly for lunch or even just a quick phone call. She always shared the current state of the company with him, as well as any ideas that she had for new initiatives. She knew that she had a unique opportunity to tap into Steve's more than three decades of experience in the camera retail industry, and she was never one to ignore such quality advice.

This week when she met Steve for lunch, she had a specific idea to share with him. For years the store had focused on generating new business through a variety of channels. Point and Shoot partnered with local universities to become the retailer of choice for their photography students. The stores sponsored local events in their communities to increase the brand reputation of the chain. Managers offered free two-hour classes for beginners with the hope that they would subsequently enroll in paid class sessions.

Although many of these initiatives had been successful in generating new customers, Cheryl lamented the fact that many customers purchased only their cameras from Point and Shoot stores and ordered their camera accessories such as camera bags, memory cards, and photo-editing software from online retailers such as Amazon or Rakuten. She wanted to entice consumers to purchase their entire suite of accessories from Point and Shoot so that the store could capture more value from each customer. New customers were difficult and costly to develop, and the chain could benefit greatly from realizing more value from each customer that it was able to generate.

To accomplish this goal, Cheryl wanted to start offering bundles of cameras along with various subsets of accessories. Point and Shoot had always simplified its pricing by selling all of its items individually, and sales of many accessories had been sluggish for several years as a result. Cheryl hoped that new camera bundles would entice many consumers to buy their accessories along with their new cameras.

When she shared this idea with Steve, he responded favorably overall. It sounded like a good idea to him; however, he cautioned Cheryl to spend some time on developing the specific product mix in the bundles as well as the price of each offering. Steve knew that consumers today are very savvy shoppers and mistakes in pricing the bundles could allow for arbitrage opportunities. Strategic consumers could purchase groups of items, some of which they do not want, and sell them to other consumers in the secondary market through outlets such as eBay or Craigslist. Steve thought that product bundles and their corresponding prices should be designed with specific consumers in mind and should provide incentives such that the consumers select the option that was designed for them when offered a menu of choices.

Steve's advice made sense to Cheryl, but she was concerned that she wouldn't be able to price the bundles effectively. When she expressed these hesitations to Steve, he suggested that she partner with a Marketing Research class at one of the local universities to have the students conduct a consulting project around the pricing of the bundles.

Market Research

To pare the consulting project down to something that the students could accomplish in a few weeks, Cheryl worked with the professor to limit the scope of the analysis to bundles of the most popular model of digital SLR camera along with a camera bag, memory card, and photo-editing software. The unit costs that the store pays its supplier for each item are provided in Exhibit 5.1.

The students developed a questionnaire that they emailed to thousands of people who had identified themselves as camera novices and in the market for a new camera. Cheryl was able to obtain this list of potential customers from an industry group. The questions were designed to identify each respondent's reservation price for each product offering, which represents the maximum amount that he or she would be willing to pay. Customers can be assumed to select the product option that provides the highest consumer surplus, the difference between the reservation price and the unit price.

Based on their analysis of the responses, the students were able to classify the responses into six types of potential customers, each with their own reservation price for the individual products as well as a bundle of all of them and a bundle that excludes the software. Each of these groups of customers is discussed in the following list. Exhibit 5.2 provides a graph of the relative size of the respondents in each customer group:

- **Professional software novice**—These customers are ready to jump into the hobby fully and want to learn how to use professional photo-editing software. They value all of the products highly.
- **Free software novice**—These customers are ready to engage in the hobby, but they expect to use free software applications such as Google Picasa to edit their photos. As a result, they do not value the photo-editing software highly.
- **Camera only**—These customers only want to purchase a camera from Point and Shoot. They want to purchase all of the other items elsewhere.
- **Memory card only**—These customers are only in the market for a memory card for an existing camera that they have purchased previously.
- **Camera bag only**—These customers are looking to replace their existing camera bag.
- **Software only**—These customers have been using free software applications to edit their photos, but they want to graduate to a professional software package.

The students recommended designing product bundles to meet each group's needs. The professional software novice group should be incentivized to purchase a bundle of all four products. The free software novice group should purchase a bundle that includes a camera, memory card, and camera bag without any software. The final four groups should want to purchase the single product that interests them. Exhibit 5.3 provides the number of respondents in each group as well as the reservation prices that they have for the four individual products and the two types of bundles.

Developing an Optimal Bundling Strategy

Cheryl appreciated the work that the students produced for her and was very impressed at their ability. Based on her knowledge of the industry, the relative breakdown of the customer types depicted in Exhibit 5.2 made sense to her, and she was comfortable making her decisions based on that assumption. The reservation prices seemed to be reasonable as well, but she had less confidence in those values due to the heterogeneity of the consumers. She wanted to incorporate some uncertainty into these estimates to develop a robust solution.

With the results of the students' report and recommendations, Cheryl was ready to develop a pricing strategy for all four products individually and for the two types of bundles. However, she knew that she had to be careful to ensure that the prices would induce each customer type to purchase the product designed for it instead of bleeding over into a different group.

Exhibits

Exhibit 5.1 Unit Costs for Products Comprising a Camera Bundle

Product	Camera	Memory Card	Camera Bag	Software
Unit cost	$525	$17	$12	$28

Exhibit 5.2 Graph of the Number of Questionnaire Respondents in Each of the Six Customer Types

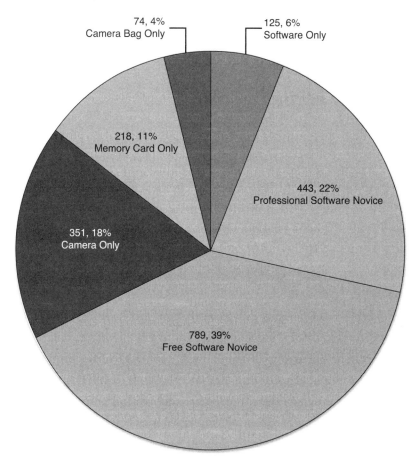

Exhibit 5.3 Reservation Prices for Each Product by Customer Type

Customer Group	Number	Camera	Memory Card	Camera Bag	Reservation Prices Software	Bundle of All	Bundle w/o Software
Professional software novice	443	800	30	20	60	910	850
Free software novice	789	750	30	20	20	820	800
Camera only	351	850	15	10	20	895	875
Memory card only	218	500	40	10	20	570	550
Camera bag only	74	500	15	30	20	565	545
Software only	125	500	15	10	75	600	525

6

Intermodal Routing for an Ocean Container

Matthew J. Drake, Duquesne University

Larry Driscoll is an international logistics specialist for Industrial Technology, a manufacturer of high-tech gas detection products for the mining industry located in Fairmont, West Virginia. Larry's main responsibility was to schedule and coordinate the logistics services for inbound shipments of materials, parts, and subcomponents from foreign suppliers around the world. Larry worked closely with his colleagues at H. P. Davidson, a licensed customs broker, who handled the documentation and insurance requirements for importing freight into the United States. Larry, however, had a great deal of experience in working with international carriers. As a result, he always made the decisions about the route that a specific shipment would take and which carriers it would travel on to get to Fairmont.

One Friday morning before Larry could even pour his usual cup of coffee, he received a call from Susan Davis, one of Industrial Technology's senior buyers. She told him that she had just finished negotiating a contract with Xinhua Electronics Company to purchase a 20-foot shipping container full of parts used to produce the company's main line of gas detectors. Susan explained that Industrial Technology's manufacturing planners had become frustrated with the company's current supplier of these parts due to repeated shipment delays and missed delivery deadlines. They suggested that Susan start looking for new suppliers that could provide better service.

Susan quickly identified Xinhua Electronics as a possible new supplier for the parts. Xinhua has developed an exceptional reputation in the industry for production quality since it had been founded five years ago. Industrial Technology had been relatively happy with its current supplier up until recently, so its procurement personnel had not thought it necessary to consider a switch to Xinhua. Now, however, the situation was different, and Xinhua looked like a great option. For all of the talk about its stellar quality, Susan had not heard much about Xinhua's logistics operations. As a result, she requested a quotation based on the Ex Works Incoterm, which made Industrial Technology responsible for coordinating and paying for all of the logistics required to transport the parts from Xinhua's manufacturing plant to Fairmont.

Xinhua Electronics planned to ship the parts from its production facility in Changsha, China, which is located southwest of Shanghai and northwest of Hong Kong. Susan told Larry that the terms of the contract with Xinhua and the letter of credit that Industrial Technology would use to pay for the goods had specified that the container would be ready to ship from Changsha by Tuesday of the following week. Industrial Technology's production schedule required that the parts arrive in Fairmont no later than 30 days from Tuesday. Any shipment delay would jeopardize Industrial Technology's ability to meet its customers' deadlines.

Company Background

Industrial Technology was founded in 1994 by Logan Richardson, a 20-year veteran of the mine safety equipment industry. Logan had spent his entire career in various sales and operations positions with a major mine safety equipment manufacturer, and he had eventually become fed up with that firm's lack of commitment to its customers that had developed due to the company's rapid expansion efforts.

Logan wanted to start a company founded on customer service and safety first.

He decided to locate the company's headquarters and manufacturing plant in the Interstate-79 High Technology Corridor that had recently opened near Fairmont, West Virginia. This area has attracted many technology firms, including IBM, Northrop Grumman, and Lockheed Martin as well as several U.S. government entities, such as NASA and the FBI. Logan felt that his employees would be inspired by the innovative spirit of the region to push the boundaries of new product development in the mine safety equipment industry.

Intermodal Routing Options

Based on his many years of experience with import shipments, Larry quickly mapped out in his head the path that the container would take from China to Fairmont. The container must first be transported to a Chinese ocean port via truck or rail. He began his analysis by checking the ocean shipping schedules at the ports near Changsha to determine which vessels would be good candidates to move the shipment based on the timing of their arrival to and departure from the ports. It looked like Shanghai and Hong Kong would be the only two feasible options for this shipment.

Once the container was loaded onto the vessel, it would be transported across the Pacific Ocean and would arrive at a port on the West Coast of the United States. Both of the potential vessels were scheduled to call on the ports of Seattle, Oakland, and Long Beach. The container would enter the United States at one of these three ports.

After the container was released by U.S. Customs at the port of entry, it would be loaded onto a rail flatcar for its long-haul transport across the country. It could travel to one of three large rail hubs: Chicago, Memphis, or Kansas City. At that point, the container would be sorted and loaded onto a different train to continue its journey eastbound. Its final rail destination could be Columbus, Louisville, or Pittsburgh. From there, the container would travel by truck to the Industrial Technology facility in Fairmont.

Larry called his contacts at his customs broker, H. P. Davidson, and together they were able to determine the cost of each potential leg of the container's journey. Because this is a relatively time-sensitive shipment, they also noted the travel time required on each route. Both sets of information are summarized in Exhibit 6.1. Larry also estimated additional handling time that would be required to transfer the shipment between modes of transportation, such as from truck to ocean vessel and from ocean vessel to rail. These handling times are provided in Exhibit 6.2. Note that these only apply to ocean ports and rail hubs. There is no significant time to transfer the container from railcar to truck in Columbus, Louisville, or Pittsburgh in preparation for its final delivery.

Route Selection Decision

Armed with the information about the possible routes and their cost, transportation times, and applicable handling times, Larry was ready to determine the lowest-cost route that was expected to meet the shipment's 30-day deadline. Although he knew that his handling time estimates, and the transportation time values for that matter, could vary, he felt comfortable using these values as fixed rather than variable in his analysis. He would be tracking the shipment regularly as it moved closer to Fairmont, so he knew that he could contact the carriers to expedite the shipment if that turned out to be necessary because it had been delayed during one leg of the journey.

Larry also knew that the 30-day deadline would probably restrict his routing choices for the shipment compared with the minimum-cost route with no firm delivery deadline. Larry wondered how much

Industrial Technology could save on the transportation costs if the delivery deadline could be extended by one or two days. Perhaps Xinhua Electronics could have the shipment ready on Monday instead of Tuesday, or maybe Industrial Technology's manufacturing planners could adjust the production schedule slightly to allow for the shipment of parts to arrive one or two days later. He decided to include an estimate of these potential savings in his analysis as well to make the best possible logistics decision for this shipment.

Exhibits

Exhibit 6.1 Transportation Cost and Time on Each Possible Route Segment

Origin	Destination	Transportation Cost	Transport Time (in Days)
Changsha	Shanghai	6,000 RMB	2
Changsha	Hong Kong	4,000 RMB	2
Shanghai	Seattle	14,000 RMB	12
Shanghai	Oakland	17,000 RMB	14
Shanghai	Long Beach	19,000 RMB	18
Hong Kong	Seattle	16,000 HKD	14
Hong Kong	Oakland	15,000 HKD	15
Hong Kong	Long Beach	18,000 HKD	16
Seattle	Chicago	1,500 USD	4
Seattle	Memphis	2,100 USD	5
Seattle	Kansas City	1,700 USD	4
Oakland	Chicago	1,800 USD	5
Oakland	Memphis	2,000 USD	5
Oakland	Kansas City	1,600 USD	4
Long Beach	Chicago	2,400 USD	6
Long Beach	Memphis	1,900 USD	5
Long Beach	Kansas City	2,200 USD	5
Chicago	Columbus	600 USD	1
Chicago	Louisville	700 USD	1
Chicago	Pittsburgh	900 USD	2

Origin	Destination	Transportation Cost	Transport Time (in Days)
Memphis	Columbus	700 USD	1
Memphis	Louisville	500 USD	1
Memphis	Pittsburgh	1,000 USD	2
Kansas City	Columbus	800 USD	2
Kansas City	Louisville	500 USD	1
Kansas City	Pittsburgh	1,100 USD	2
Columbus	Fairmont	400 USD	1
Louisville	Fairmont	700 USD	2
Pittsburgh	Fairmont	200 USD	1

Exhibit 6.2 Handling Time (in Days) for a Container at Each Transshipment Location

City	Handling Time (in Days)
Shanghai	3
Hong Kong	2
Seattle	5
Oakland	4
Long Beach	3
Chicago	2
Memphis	2
Kansas City	3

Part II
Analytics in the Service and Utility Industries

7 Lolly's Restaurant 77

8 Strategies for Managing Service Delivery Gaps and
   Service Recovery 83

9 EverClean Energy, Inc.: Wind Energy Versus
   Natural Gas 95

10 Slotting Pharmaceuticals in an Automated Dispensing
   Cabinet. 111

7

Lolly's Restaurant[1]

Steven S. Harrod, Technical University of Denmark

Roger unlocked the door and entered the familiar scent of donuts, hamburgers, and coffee that was the trademark of Lolly's Restaurant. As the light flickered on and the employees trickled in, Roger made his morning inspection before the breakfast rush. Roger greeted Gwen, the hostess, and made sure that the breakfast menus were in place at the podium.

Lolly's served two meals only, breakfast and lunch. Opening at 6:00 a.m., Lolly's was ideally located on a heavily traveled commute route toward downtown Lancaster, Pennsylvania. Patrons included Lolly's regionally known donuts and coffee in their morning commute, and Lolly's also enjoyed a respectable lunch traffic for hamburgers and sandwiches. Lolly's remained open during the off-peak hours between 9:00 a.m. and 11:00 a.m., but it was very quiet.

Earl Sweat opened Lolly's in 1963, starting first with a roadside stand and curbside service. "Lolly" was his daughter Annette's favorite doll. From that roadside stand, Lolly's had grown to a table restaurant that seated 125 persons. The menu consisted of American comfort food: eggs, meats, cereals, coffee, pastries, hamburgers, fries, deli sandwiches, and soup. The typical customer check was $12, and Lolly's was collecting approximately $6,000 per day in revenue, evenly divided between breakfast and lunch. Lolly's closed at 2:00 p.m. after lunch, and was closed on Sundays.

Earl was elderly, approaching 90 years old, and daily management had passed to Roger Engeman years ago. Earl was making plans to sell shares of the restaurant to local business investors as part of his family estate plan. Many of these potential investors had asked why Lolly's did not open for dinner. Earl, with Roger's assistance, was considering a dinner service plan as a way to entice local investors to buy shares of Lolly's. If Earl could sell 70% of the equity in Lolly's, he felt he could finalize his estate plan for his 16 grandchildren and finally be at peace with his life's goals.

Lolly's was debt free. Earl owned the property outright, consisting of a 3,000-square-foot, single-story restaurant and a 100-space parking lot, with 300 feet of frontage on Lincoln Highway. (See Exhibit 7.1 for a map of Lolly's physical location.) Property taxes were approximately $44,000 per year, and insurance was an additional $12,000 per year. The primary utility expense was electricity, at about $3,200 per month. Lolly's had an advertising budget of $5,800 per month. Lolly's spent half a million dollars on groceries and restaurant supplies last year.

Roger had performed some field research on comparable restaurants. With the assistance of some paid "spies," Roger had collected car traffic data for other Lancaster full-service restaurants and found that on average car traffic during the peak two hours of dinner was about 60% of the traffic level during the peak two hours of lunch. Most restaurants that opened for dinner closed at about 10:00 p.m.

Roger then turned to the operations of the restaurant. Waitstaff earned $5 per hour, and kept their own tips. Payroll benefits and taxes cost Lolly's an additional 30%. Waitstaff cleaned and bussed their own tables. Kitchen staff earned an average of $14 per hour, depending on skill and seniority. Gwen, the hostess, earned $12 per hour, and Roger, the manager, was paid a salary of $36,000 per year.

The restaurant floor contained 29 tables, most of which seated four persons, but a few seated less. A counter seated 14 persons. Normally, six waitstaff could cover all of the tables and the counter, and six chefs worked in the kitchen. During peak periods, work in the kitchen was like a game of whack-a-mole. Chefs cooked multiple meals simultaneously, placing raw eggs and meat on the grill one second and then scooping up adjacent cooked items the next. Some items, such as soups and chili, were prepared in advance during off-peak hours, but even these required some time to ladle and arrange with side dishes at serving time.

Roger compiled a table of popular menu items and the time to prepare them in the kitchen (Exhibit 7.2). "Lead Time" refers to how long an entrée requires to deliver after request (cooking time), and "Labor" refers to how much payroll time is allocated as cost to that entrée, including advance preparation time. Times are valid under steady conditions where kitchen orders are evenly distributed, and do not include any delays due to fluctuations in customer demand.

Roger then also consulted with a master chef to estimate similar times for projected dinner menu items (Exhibit 7.3). Roger hoped to price the menu so that the average check was $16. With the time available, Roger thought this was the best he could do in research and data collection, and it was adequate to make his analysis.

The whole decision made Roger a little nervous, not the least because he wondered if his reward for all this work would be to lose his job under the new owners. Why rock the boat? Why not just keep Lolly's operating as it had been and be happy with the steady income? Roger collected his thoughts and mentally noted what he needed to do. First, he needed to estimate the cost of operation for the dinner service. Second, he needed to estimate the potential net revenue. Third, he would have to estimate the start-up costs. How long would it take to earn back the cost of launching the new service?

All of that sounded straightforward, but there was more. What was the risk or variability in his data? What if his numbers were wrong? What about restaurant operations? Would dinner operate the same way as lunch (or breakfast)? Would Roger's job be safer if he advised against expansion? Roger planned to use the break between meals today to review the collected data and deliver a recommendation to Mr. Sweat.

Study Questions

1. How many employees must be hired to work the expanded schedule?
2. What are the expected revenue and cost of goods sold for the new meal service?
3. What is necessary to judge the expanded hours a success?
4. Will customer service during the expanded hours be better or worse than normal breakfast and lunch service?

Endnote

1. The characters and firms depicted in this case are fictional, and any similarity to real past or present persons is entirely coincidental.

Exhibits

Exhibit 7.1 Physical Location of Lolly's Restaurant in Lancaster, Pennsylvania

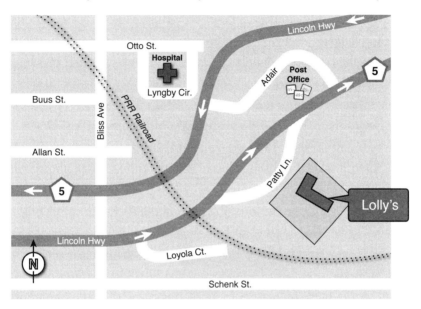

Exhibit 7.2 Sample Kitchen Times for the Current Menu

Breakfast	Lead Time (in min.)	Labor (in min.)	Lunch	Lead Time (in min.)	Labor (in min.)
Steak and eggs	9.2	2.1	Club sandwich platter	5.2	2.9
Eggs platter	7.9	1.9	Chili and salad	4.9	2.9
Oatmeal platter	7.2	2.3	Cheeseburger platter	6.9	2.5
Biscuits and gravy	6	1.7	Soup and salad	4.5	1.4
Pancake platter	9	2.6	Grilled cheese platter	5.4	1.8
Waffle platter	8.4	2.4	Chopped steak	8.6	2.6

Exhibit 7.3 Sample Kitchen Times for the Proposed Dinner Menu

Dinner	Lead Time (in min.)	Labor (in min.)
Angus burger platter	7	4.7
Rib eye platter	11.6	4.9
Lasagna and salad	9.2	3.8
Irish stew	5.4	2.7
Turkey platter	7.5	4.9
Fish sandwich	6.4	4.7

8

Strategies for Managing Service Delivery Gaps and Service Recovery

Ajith K. Thomas, Saintgits Institute of Management, Kerala, India

The hotel rooms in Bangalore, India, were running full due to the annual air show event at the parade grounds. SM International, a four-star luxury hotel mostly patronized by corporate middle management executives, was running at minus occupancies. Minus occupancies occur when the hotel has a certain number of rooms to sell and has confirmed bookings for a higher number of rooms than is available at the hotel.

SM International has 94 rooms, and the reservations on the books were for 103 rooms for that day. This meant that they had to arrange for nine more rooms by the end of the day, as all bookings had been reconfirmed with those guests and bookers whose names are on the arrival list. Being a business hotel located in the center of the city, traditionally they always did some overbookings in anticipation of some last-minute cancellations.

It is quarter past one in the afternoon, and guests for tonight have already started checking in. The revenue manager and front office manager (FOM) had a meeting earlier today, managing to shift the low-paying and discount clientele (usually corporate guests) to competing hotels in the area where they had booked some rooms for the night at a lower rate. Guests staying at their hotel the night before were reminded, very diplomatically, that their checkout time was 12:00 p.m. in anticipation of the hotel running a full house tonight.

The challenge for them was to allocate the remaining guests between the two different categories of rooms the hotel offered—the 16 suites and 78 deluxe rooms. A few regular guests and important people staying at the hotel were upgraded to the suites. As a strategy to maximize revenue, the revenue manager suggested that the FOM request guests to double up for some rooms if the guests were from the same company. From their earlier experiences, they know that because the hotel is commonly patronized by middle-level managers, there are greater chances that they would share rooms if the hotel offered them complimentary cocktails. Every guest from the same company who was approached by the FOM fortunately agreed to this idea, especially when the hotel informed them that this arrangement was primarily due to the scarcity of rooms in the city. The hotel promised every guest who wished to shift back to the SM International hotel that they could come back the next morning after breakfast, as they were anticipating some checkouts around that time.

Background on the Hotel Industry in India

The hotel industry is a significant stakeholder in the Indian tourism sector. There are more than 1,250 hotels approved and classified by the Ministry of Tourism, with a total capacity of more than 75,000 hotel rooms.[1] The Indian hotel industry is adding approximately 40,000 four- and five-star category rooms in the major cities in the next few years.[2] The hotel industry is fragmented, seasonal, and labor intensive. Hotels in India generally have low occupancy levels during the monsoon months, and the period from October to April has the highest occupancy levels. In India, 60% of the annual hotel business occurs between December and March. The hotel industry employs skilled, semiskilled, and nonskilled labor directly and indirectly. The fragmentation of the industry in the unorganized and organized sector can be seen reflected in the Herfindahl–Hirschman Index, a commonly accepted concentration ratio of firms.[3] In a hotel, the revenue generated from room rates is often greater than 70%, while that generated from other facilities, such as restaurants and bars, is less than 30%. A greater number of rooms occupied leads to a larger number of guests in-house and an increased opportunity for more guests to use the revenue-generating hotel facilities, which may include room service, the business center, banqueting, conference facilities, and restaurants. Thus, higher room occupancy is a key to better performance of any hotel.

Although it is true that the demand for hotel rooms is often much greater than supply, the seasonal nature of the business requires focused marketing to keep the rooms filled. Therefore, the selling of rooms is crucial to the performance of any hotel. Hotel rooms are used by leisure travelers, business travelers, air crew, banquet guests, and free individual travelers (FIT). They can be recognized as different segments of the business for tracking the revenue of any hotel. Out of all of these segments, the business travelers account for more than 60% of room occupancy, especially in hotels that are in the four- and five-star category.

Large hotel operators such as Hilton, Four Seasons, Ritz Carlton, Starwood, Hyatt, Marriott, Intercontinental, Taj, Oberoi, and Leela are in expansion mode in India, and most investments are in the five-star deluxe category, the five-star category, and the budget category, a term that applies predominantly to the business travelers. Due to the demographic divide, high GDP growth, and FDI prospectus, business growth in India is positive. However, this growth and the gap between the demand and supply cannot be taken as an excuse not to focus on the room sales especially because of the seasonality of the business, which was discussed above. This can be further explained as follows. If a hotel X has 100 rooms, it is possible to sell 35,000 rooms in a year, leaving 15 days or so for maintenance. However, seasonal variations actually lead to far fewer room sales per year. In other words, the fixed cost of the rooms cannot be appropriated over 35,000 nights unless the sales force keeps the rooms filled. While there is pull demand during the season, push demand is an important aspect of sales during the off-season. Therefore, hotels undergo a highly fluctuating demand-supply cycle depending on the season. It is here that the sales of rooms to corporate clients are most significant, as business visitors use the rooms based on business requirements only and not based on the season, as the conventional tourists do.

India is going to be a significant individual contributor to the global demographic transition through the year 2050.[4] The growth experienced in India in the 1980s and 1990s is at least partially attributable to the country's changing demographics and its economic reforms.[5] Some analysts have predicted that India will have larger annual rates of GDP growth than China through 2025, and its economy at that point will likely be larger than all European countries and similar in size to that of Japan.[6] This means that more industries are likely to enter India to take advantage of growth opportunities, and the demand for hotel rooms for corporate use will likely increase even further. Leisure demand will likely increase as well with the Indian government's recent move to expand the ability for travelers from 43 countries (up from only 12) to collect their Indian travel visas upon arrival at one of nine international airports in India instead of having to acquire them before their initial departure.[7]

Service Failure

Checkouts from the previous night and check-in for the current night went very smoothly until late at night when there was only one room left and two guests were expected to arrive. They were from a reputable software company that has been giving good business to the hotel. The FOM checked the names of the guests and confirmed with the local office that two management trainees were expected from their Korean office that evening around 6:00 p.m. Because the names were Korean names, the local room booking personnel (booker) spelled out the guest names for the FOM. The FOM informed the booker about the fact that they may have to share a room, as the hotel was running full. The booker did not agree to this arrangement and asked the FOM to check with the guests about the sharing possibilities when they arrived. He also added that if the guests did not have a problem with sharing, he would not have a problem with them doing so.

One of the two guests arrived at the hotel exactly at 6:00 p.m. that evening, and the FOM told him that the hotel had proposed a twin-sharing arrangement for them due to the lack of available rooms. He asked the guest whether he would like to shift himself to a nearby hotel where there was a room available at a lower rate. The guest declined this offer and said that he would share with the person coming from his own company, as he felt comfortable to stay with the hotel brand. He even showed the FOM the membership card for the hotel brand, reiterating the fact that he is a regular guest with the same brand back in his home country.

The FOM marked the check-in of the guest in the system, and his luggage was sent to the room. The guest was personally escorted by the FOM to the room, and the FOM also asked the guest to join him for dinner that evening in the restaurant around 8:00 p.m. Because it is a regular practice for the hotel to check in both of the guests at the same time for all double occupancies, he showed the room in the system as double occupancy and left a note for his reliever in the next shift (who will come in at 8:00 p.m.) that he had convinced the guest to agree to the double occupancy and he was ready to share the room.

Time went by, and at about 3:00 a.m. the other guest for the room finally arrived at the lobby. The person looked very tired as if the journey had been very difficult, and jet lag was very evident. There was a flight cancellation, a delay, and the guest had traveled for almost 24 hours from the time the journey had started. The check-in was very smooth, and the guest was escorted by a front office executive to the room on the ninth floor with the room key.

The executive followed the standard procedure of knocking three times on the door, opened the door, kept luggage on the luggage rack, and escorted the guest to the bed. Suddenly, to his surprise, he saw a man on the bed. The female guest who had just been escorted to the room screamed after seeing the man on the bed. The man on the bed was trying to convince the other guest that he had checked in earlier, and she seemed to have understood nothing. She was perplexed and neither understood what the executive was saying nor what the man on the bed was saying.

A minute went by, and it seemed like an hour for all three in the room. Recouping from the initial seconds of the incident, the man still on the bed started thinking, "A girl sharing my hotel room! What will I tell my wife, when I get back home? I already told her over the phone that there was a sharer tonight, and she joked that she hoped that it was not a female guest." In the back of his mind, he had remembered that the bed in the room was a queen-sized bed, which he wanted to remind the FOM while they met at the restaurant for dinner but somehow missed telling him.

How should the hotel manage the current situation? The booker will likely come to know of this bungle early the next morning and is sure to react. He may even say that they will never use such an unprofessional hotel in the future. What will the customer service team do to recover the business due to this service failure? Can they face the booker with ease? What will the hotel management do so that these errors never get repeated?

Service Recovery

In the aftermath of the late-night check-in issue, the hotel staff and management undertook the following actions to handle the incident for the affected parties as well as to improve the service offered in the future:

1. The hotel had a managing director's room, which is usually occupied only by the hotel owner when he is in town. The room is fully furnished and quite similar to the executive rooms of the hotel. This was never sold to guests and was not shown in the inventory of rooms. This room was immediately rechecked, and the female guest was shifted to this room. The manager on duty who was staying in the hotel was immediately notified about the incident, and he quickly met the guest and apologized to her. Because the situation was handled very diplomatically over a coffee at the lobby, the guest also was satisfied with the resolution. She especially commended the manager on duty for coming to her rescue and handling the situation even in the early hours of the morning.

2. The next morning, the sales manager and FOM walked across the street to the company where they met the booker and briefed him of the situation and how it was handled. This seems to have boosted the confidence he had in the hotel.

3. At the direction of the FOM, a front office team conducted a service blueprinting of the check-in procedure, where key issues were marked as (F) to identify a possible failure point in the process. Blueprinting enables the service provider to carefully manage the service delivery process. From a customer perspective, the most serious fail points are marked as (F) whose failure will inhibit their ability to enjoy the core product.

Service Process Improvement

This incident has left a strong impression on the FOM. He is comfortable that he has taken measures to correct this particular issue with the check-in process so that this service failure is unlikely to occur again. He realizes, though, that there are other parts of his hotel's operations that are not perfect and are likely to result in service failures in the future. He has resolved to begin an initiative that will examine the most commonly occurring service failures, determine their underlying causes, and develop recommendations to improve the process to limit future failures.

Service failures in the hotel can be broadly classified as (1) room-related complaints, (2) food- and beverage-related complaints, and (3) other service-related complaints. The FOM has collected data on the frequency that each service failure occurred over the past two years (2013–2014), which is provided in Exhibit 8.1. Until now, the FOM had only collected the data but had not acted upon it. This recent service failure has illustrated the importance of providing quality service to every customer, and the FOM is committed to improving all aspects of the hotel's service operations.

The list of service failures in Exhibit 8.1 is quite long, and the FOM knew that he could not try to address all of them at once. He had to prioritize the failures to determine which ones to concentrate on first. This prioritization would also provide guidance to the other hotel employees who will be required to execute many of the process modifications. How can the FOM sequence the service failures in his process improvement strategy?

Conclusion

The service recovery paradox represents the positive feeling that the customer experiences after the resolution of a service failure. These customers are often far happier to use the service provider in the future than customers who had no service problems. As a result, customer recovery should be seen as a profit center and not as an expense center. An unsatisfied customer may feel unhappy and may tell many of his friends, thereby damaging a future stream of business for the company. Recovery of customers from service failures represents the organization's pious expression of determination to resolve the situation. It requires commitment, planning, and clear guidelines on the part of the organization; and the recovery procedures should be proactive, structured, planned, and implemented by empowered and trained employees.

Endnotes

1. Ministry of Tourism, Government of India, "India Tourism Statistics," http://tourism.gov.in/writereaddata/CMSPagePicture/file/marketresearch/publications/INDIA%20TOURISM%20STATISTICS%202013.pdf.

2. Knight Frank Research, "India Hotel Market: Introspection and Outlook," http://content.knightfrank.com/research/318/documents/en/december-2010-403.pdf.

3. Dun & Bradstreet India, "Indian Hospitality Industry," https://www.dnb.co.in/Travel_Tourism/Indian_Hospitality_Industry.asp.

4. J. Bongaarts, "Human Population Growth and the Demographic Transition," *Philosophical Transactions of the Royal Society B: Biological Sciences* 364(1532): 2985–2990.

5. A. Virmani, "Accelerating and Sustaining Growth: Economic and Political Lessons" (IMF working paper WP/12/185) http://www.imf.org/external/pubs/ft/wp/2012/wp12185.pdf.

6. J. O'Neill, "Winners and Losers in the Global Economy of 2025," *Europe's World* (Autumn 2014), http://europesworld.org/2014/10/09/winners-and-losers-in-the-global-economy-of-2025.

7. World Travel & Tourism Council, "Easing Restrictions—What Will India's New Visa Stance Mean for Tourism?" http://www.wttc.org/global-news/articles/2014/dec/easing-restrictions-what-will-indias-new-visa-stance-mean-for-tourism/.

Additional Readings

C. H. Lovelock, P. G. Patterson, and R. Walker, *Services Marketing: An Asia-Pacific and Australian Perspective*, 4th ed. (Sydney: Prentice Hall Australia, 2007).

C. H. Lovelock, J. Wirtz, and J. Chatterjee, *Services Marketing: People, Technology, Strategy*, 7th ed. (Upper Saddle River, NJ: Pearson, 2011).

Exhibit

Exhibit 8.1 Yearly Service Recovery Instances at SM International for the Period 2013–2014

		Frequency	Percentage
1	**Room-Related Complaints**	**81**	
a	Luggage transfer	8	9.88
b	Air-conditioning	18	22.22
c	Room allotment wait	11	13.58
d	Newspaper delivery delay	6	7.41
e	Wake-up calls	5	6.17
f	Room amenities failure	12	14.81
g	Key card programming failure	14	17.28
h	Sound from other rooms	7	8.64
2	**Food- and Beverage-Related Complaints**	**69**	
a	Wrong order delivery	7	10.14
b	Delay in service	16	23.19
c	Cold food service	9	13.04
d	Beverage quality complaints	11	15.94
e	Service personnel related	10	14.49
f	Food quality complaints	10	14.49
g	Buffet replenishing delays	6	8.70
3	**Other Service-Related Complaints**	**84**	
a	Transaction delays	9	10.71
b	Taxi support issues	14	16.67
c	Card machine complaints	13	15.48
d	Billing issues	8	9.52
e	Wi-Fi complaints	15	17.86
f	Housekeeping issues	9	10.71
g	Laundry problems	16	19.05
	TOTAL	**234**	

9

EverClean Energy, Inc.: Wind Energy Versus Natural Gas

Patrick Cellie and Matthew J. Drake, Duquesne University

Introduction

Andrew James was admiring the wind turbines his team had just finished installing after months of hard work. Even though he was the CEO of EverClean Energy, he still enjoyed visiting the sites where the company was developing new projects. He had more than 20 years of experience in renewable energy, but for the first time he didn't have the same confidence he had always had about his company's future growth potential. Natural gas prices had dropped to levels that made it difficult for renewable energy sources to be competitive, and Andrew was afraid that in the future this situation would start affecting his business.

Observing the turbines' blades rotate in the afternoon light, he was trying to clear his mind and think about what strategy EverClean Energy should pursue in order to stay competitive in the market and keep growing. Would the natural gas prices increase again? Could EverClean afford the risk of waiting to see it happen? What alternatives did the company have? He needed to come up with ideas to discuss in the following week's board meeting.

Company Background

Andrew James founded EverClean Energy, Inc., in 2002 with a focus on developing cost-effective methods for wind energy production. He believed the effects of climate change, including the destruction caused by recent natural disasters, such as Hurricane Katrina and Superstorm Sandy, provided compelling reasons why the United States and the world must have a dependable and reliable source of clean energy. In Andrew's vision, wind energy currently was the only economically viable and scalable form of renewable energy. It is local and reliable, displaces air pollution, and is not subject to volatile fuel markets.

The company had been successful so far. In nearly a decade, seven wind farms capable of producing a total of 750 megawatts (MW) had been installed and were operating, while an additional 1,400 MW were under development. Most of the projects were in the Mid-Atlantic states, such as Pennsylvania, New York, Ohio, and West Virginia, but recently the company had started working in other areas, such as Illinois, Montana, and California.

Revenues had constantly grown over the decade, and from 2009 to 2012 they had gone from $4 million to $42 million. Forecasts for the following years were optimistic, but the recent drop in natural gas prices represented a threat.

The company believed in a triple bottom line approach and was proud of the impact it was having in terms of environmental and social sustainability. Through the implementation of its projects, EverClean Energy had avoided a total of about 720,000 tons of GHG emissions and produced energy for an equivalent of 120,500 households through 2012. The company was also very proud of its involvement with the communities that resided in the areas where the projects were developed. For farmers and landowners, wind turbines provided supplementary income, while communities benefited from the tax revenue that supported schools, hospitals, road improvements, and other projects. More than 450 jobs were created during construction of the turbines, and investments in local goods and supplies of more than $21.5 million were estimated during the projects' lives.

As a wind energy provider operating in the Mid-Atlantic and Northwest regions of the United States, EverClean Energy generated business by providing a renewable energy option through the electric grid. EverClean Energy marketed its energy through electricity distributors using long-term power purchase agreements (PPAs) and other sales mechanisms. As the company continues to identify growth opportunities, it is faced with a number of competitive challenges. Many states have renewable portfolio standards (RPS) that will continue to support growth in clean energy development, but ideal locations for wind farms are becoming scarce, which creates difficulty in achieving desired rates of return on projects. In addition to competing with other wind energy developers, EverClean Energy will need to compete against other sources of electricity to remain viable beyond RPS minimums and continue to grow the company.

Wind Energy

Onshore wind turbines are one of the lowest-cost methods of generating electricity. Electricity from wind turbines requires no fuel cost, which has enabled producers to offer their customers fixed electricity rates that last for 20–30 years via contracts known as power purchase agreements (PPAs). The Department of Energy's Wind Technologies Market Report suggests that these PPAs are a result of the reduction in the cost of generating electricity from wind power over the past few years.

These recent cost reductions have allowed producers to offer consumers more affordable rates through the PPA contracts. In 2012, PPAs signed for electricity generated by wind turbines were between $31 and $84 per MWh (megawatt-hour). The large spread in the range of prices is due to the location and size of the installation of the turbines. This is a significant drop in the rate from 2010, which was between $44 and $99 per MWh.[1]

Drivers of Wind Energy Costs

The cost of electricity generated by wind energy is determined by several main factors: capital investment costs, capacity availability and efficiency, and ongoing operating costs. As the technology used in designing wind turbines improves, manufacturers are able to build bigger turbines that have higher capacity and efficiency. These technological innovations lead to lower-cost electricity production.[2]

Capital Investment Costs

The wind turbine equipment itself often represents the majority (as much as 70% or more) of the cost of investing in an on-shore wind energy project. The remainder of the costs are related to the installation and construction of the turbines. Technological advancements in recent years have reduced the overall capital costs of wind projects by 10–15% compared with the costs in 2010.[3]

Capacity Availability and Efficiency

The strength and consistency of the wind that is available at a wind farm site will impact the costs of the wind electricity by determining how much energy a wind project can produce. Typical capacity factors, which represent the percentage of theoretical capacity that an installed turbine can realize, are 15% to 50%; values at the upper end of the range are achieved in favorable sites and are due to wind turbine design improvements. More advanced technology, such as taller towers and larger and lighter blades on the turbines, has allowed firms to capture wind more efficiently. These advancements improve the performance of the turbines by increasing the capacity factor.[4]

Ongoing Operating Costs

Once a wind project is implemented and starts producing energy and providing it to the power grid, it has an expected life of at least 20 years. Over the course of its useful life, a wind turbine must be monitored and maintained by wind technicians and operators to keep a high level of performance. For modern machines, the estimated maintenance costs are in the range of 1.5% to 2% of the original investment per year, which is down from the 3% that was required in the 1980s.[5]

Other

The cost of the electricity generated by a wind turbine implementation is also affected by other factors, including the cost of financing the equipment; specific costs related to the site of the wind farm, such as building and operating permits; the cost and potential scarcity of specialized skilled labor; and the cost of transportation and logistics.[6]

Incentives, tax credits, and regulations such as carbon taxes also play an important role in the definition of the price. In the last few years in the United States, a 30% tax credit and renewable energy credits (RECs) have supported the growth of the industry.

The Declining Cost of Wind Energy

Technological improvements and the cumulative effect of learning in the industry have facilitated a significant reduction in the cost of electricity produced by wind power over the past three decades. The U.S. Department of Energy estimates the current cost of wind energy to be between $80 and $120 per MWh. This represents a decrease of more than 80% from its value of approximately $550 per MWh in the 1980s.[7]

Over a 20-year useful life of a wind energy project, the dynamics of the cost of wind energy are extremely uncertain. It will likely continue on a downward trajectory, but the exact path cannot be determined in advance. Exhibit 9.1 provides several factors that affect the cost of wind energy generation and an estimated range of the annual impact for each factor (i.e., maximum impact, most likely impact, and minimum impact).

The last factor listed in Exhibit 9.1, the social cost of carbon (SCC), has a 25% chance of occurring in any period. Once it occurs in a period, it is assumed to occur in all subsequent periods over the planning horizon. The impact of SCC on the cost of wind power varies each year between –14.0% and –20.0%.

Factors Affecting the Cost to Generate Electricity from Natural Gas

Natural gas is an energy source that has garnered a lot of attention recently due to increased supply with the access to shale gas sources. Growth in natural gas supply has shifted the demand for energy as its price has dropped to $2.10 per British thermal unit (Btu).[8] While wind energy prices continue to decline from efficiencies and economies of scale, they are not price competitive with the depressed natural gas prices. Projections for the future of natural gas vary greatly based on a wide range of variables on the demand and supply side. A further study of these variables can help to give a better understanding of the future of natural gas.

The current market price for wind-powered electricity is around $0.09 per kWh.[9] The price is expected to decline as technologies and economies of scale continue to improve. Despite the steady decline in prices of wind energy, there is still a large gap between wind energy prices and natural gas prices, which currently stand at $0.014 per kWh after the gas is converted from Btu to kWh.[10]

Although it is difficult to predict where prices for natural gas will head in coming years, increasing supply of the resource has flooded the market with improved technological advances and regulatory approvals that have allowed for the extraction of shale gas from the Marcellus and Utica Shale, mostly located in Pennsylvania, Maryland, West Virginia, New York, and Ohio.

The increased supply of natural gas allows for many opportunities for exploitation of the resource that can otherwise spur demand and reach price parity with wind energy. The Energy Information Administration (EIA) currently forecasts natural gas prices to grow to about $6.00 Btu by 2025 and $7.00 by 2035.[11] Experts in the energy industry agree that attempting to project energy prices beyond 12–18 months is futile. Instead, experts suggest to study the issues that can affect long-term natural gas prices on both the supply and demand side.

Variables that influence natural gas prices come from both the demand and supply side. Although the number of variables that play a role in energy pricing is nearly infinite, the most relevant variables for both the demand and supply have been summarized as follows:

- Variables affecting demand:
  - Conversion of vehicles to natural gas
  - Industrial use of natural gas
  - Exportation of natural gas
  - Conversion of coal plants to natural gas
- Variables affecting supply:
  - Shale gas supply
  - Technological advances
  - Natural disasters

- Weather
- Policies

Exhibit 9.2 discusses each of these factors in more detail. Exhibit 9.3 provides examples of technological advancements in natural gas exploration and extraction that would also likely affect the cost of generating electricity from natural gas. Exhibit 9.4 suggests how some of the factors could vary and influence the cost of natural gas power. As with the occurrence of the SCC in the wind power case, once the carbon taxes and the impact fees occur in one period, they are assumed to occur in every other period in the planning horizon. The actual impact differs each year according to the range of possible values listed in Exhibit 9.4. Currently the cost of electricity produced from natural gas is between $50 and $70 per MWh.

Conclusion

Andrew James had the report with the data concerning wind energy and natural gas prices in his hand and was immersed in his thoughts. He was wondering how long the price of natural gas would remain so low and also how quickly the price of wind energy, if ever, would become comparable to it.

He was trying to list all the options he had in front of him. Should he believe it was a momentary situation and go with the flow? Or should he develop alternative business models in order to diversify EverClean Energy's investments? Should some of these options include natural gas?

He decided to approach the problem using a modeling tool that would allow him to see what the future cost of electricity generated by wind and natural gas could possibly look like years ahead by modifying the underlying assumptions and the value of some of the variables affecting the cost of the two technologies. He remembered something he learned in business school: Forecasts are always wrong, but they can still give you guidance in making a decision by providing some information about the size and behavior of the uncertainty.

Acknowledgments

This case is based on a consulting project conducted by MBA students Maureen Coyle, J. P. Gibbons, Ashley Jones, Rudy Molero, and Emily Peterson at Duquesne University in the summer of 2012.

Endnotes

1. American Wind Energy Association, "The Cost of Wind Energy in the U.S.," http://www.awea.org/Resources/Content.aspx?ItemNumber=5547.

2. American Wind Energy Association, "The Cost of Wind Energy in the U.S.," http://www.awea.org/Resources/Content.aspx?ItemNumber=5547.

3. American Wind Energy Association, "The Cost of Wind Energy in the U.S.," http://www.awea.org/Resources/Content.aspx?ItemNumber=5547.

4. American Wind Energy Association, "The Cost of Wind Energy in the U.S.," http://www.awea.org/Resources/Content.aspx?ItemNumber=5547.

5. American Wind Energy Association, "The Cost of Wind Energy in the U.S.," http://www.awea.org/Resources/Content.aspx?ItemNumber=5547.

6. American Wind Energy Association, "The Cost of Wind Energy in the U.S.," http://www.awea.org/Resources/Content.aspx?ItemNumber=5547.

7. American Wind Energy Association, "The Cost of Wind Energy in the U.S.," http://www.awea.org/Resources/Content.aspx?ItemNumber=5547.

   U.S. DOE, "Revolution Now: The Future Arrives for Four Clean Energy Technologies," September 2013, http://energy.gov/sites/prod/files/2013/09/f2/Revolution%20Now%20--%20The%20Future%20Arrives%20for%20Four%20Clean%20Energy%20Technologies.pdf.

8. CME Group, "Henry Hub Natural Gas Futures," http://www.cmegroup.com/trading/energy/natural-gas/natural-gas.html.

9. John Hanger, "New Coal Plants Now More Expensive Than New Renewables," *John Hanger's Facts of the Day*, February 27, 2012, http://johnhanger.blogspot.com/2012/02/new-coal-plants-now-more-expensive-than.html.

10. U.S. Energy Information Administration, "Henry Hub Natural Gas Spot Prices Fell about 9% in 2011," *Today in Energy*, January 10, 2012, http://www.eia.gov/todayinenergy/detail.cfm?id=4510.

11. U.S. Energy Information Administration, "Henry Hub Natural Gas Spot Prices Fell about 9% in 2011," *Today in Energy*, January 10, 2012, http://www.eia.gov/todayinenergy/detail.cfm?id=4510.

12. Energy Information Administration, "AEO2012 Early Release Overview," February 14, 2012, http://www.eia.gov/forecasts/aeo/er/pdf/0383er(2012).pdf.

13. Organisation for Economic Co-Operation and Development, "Labour Productivity Growth in the Total Economy," http://stats.oecd.org/Index.aspx?DatasetCode=PDYGTH.

14. Aaron Studwell, "Weather Patterns and Natural Gas Price Movement Linked, Trading Strategies Identified," *Weather Insight*, April 11, 2012, http://www.docstoc.com/docs/42344636/

Weather-Patterns-and-Natural-Gas-Price-Movement-Linked-Trading.

15. U.S. Department of Energy, "Energy Department Approves Gulf Coast Exports of Liquefied Natural Gas," May 20, 2011, http://energy.gov/articles/energy-department-approves-gulf-coast-exports-liquefied-natural-gas.

16. "Natural Gas Exports Offer Much to the U.S. Economy," *Washington Post*, March 14, 2012, http://www.washingtonpost.com/opinions/natural-gas-exports-offer-much-to-the-us-economy/2012/03/13/gIQA4WibCS_story.html.

17. E. Worrell, D. Phylipsen, D. Einstein, and N. Martin, "Energy Use and Energy Intensity of the U.S. Chemical Industry," Lawrence Berkley National Laboratory. LBNL-44314. http://ateam.lbl.gov/PUBS/doc/LBNL-44314.pdf.

18. U.S. Energy Information Administration (EIA), "Energy Information Administration - Manufacturing Energy Consumption Survey," http://205.254.135.7/emeu/mecs/contents.html.

19. Stuart Hampton, "Cheap US Shale Gas Spurs Plastics Boom," *Bizmology*, March 12, 2012, http://bizmology.hoovers.com/cheap-us-shale-gas-spurs-plastics-boom/.

20. U.S Energy Information Administration, "Total Electric Power Industry," *Electric Power Monthly*, March 27, 2012, http://205.254.135.24/electricity/monthly/.

21. Associated Press, "Coal-Fired Power Plants Closing: FirstEnergy Shutting Down 6 Sites in Ohio, Pennsylvania and Maryland," *Huffington Post*, January 26, 2012, http://www.huffingtonpost.com/2012/01/26/coal-power-plants-closing-firstenergy_n_1234611.html.

22. M. Spencer Green, "End of Coal Power Plants? EPA Proposes New Rules," *NBC News*, March 27, 2012, http://usnews.nbcnews.com/_news/2012/03/27/10886373-end-of-coal-power-plants-epa-proposes-new-rules.

23. "Natural Gas and Technology," *NaturalGas.org*, http://www.naturalgas.org/environment/technology.asp.

24. R. L. Mazza, "Liquid-Free Stimulations—CO_2\Sand Dry-Frac," *Petroleum Consulting Services* (2010), http://www.netl.doe.gov/KMD/cds/Disk28/NG10-5.PDF.

Exhibits

Exhibit 9.1 Summary of Possible Annual Impact of the Variables Affecting the Cost of Wind Power Generation

Variable	Maximum Impact	Most Likely Impact	Minimum Impact
Wind power price trend	6.5%	–3.0%	–8.0%
Variable operating and maintenance costs	14.5%	5.0%	–12.5%
Capacity factor	7.0%	–1.5%	–6.0%
Insurance cost	2.5%	1.5%	1.0%
Social cost of carbon (SCC)	–14.0%	–17.0%	–20.0%
Occurrence of SCC	N/A	25.0%	N/A

Exhibit 9.2 Factors Affecting Natural Gas Prices

Supply Side	Implications
Marcellus shale and future gas production	A major factor driving the low price of natural gas is the increased shale supply from the Marcellus and Utica shale. The EIA's 2012 report indicated that natural gas production will increase by 5%, whereas shale gas production will increase roughly 26% by 2035.[12] If these projections are correct and consumption of natural gas continues at the current rate, only 40% of the available shale gas will have been exhausted by 2035.

Supply Side	Implications
Environmental impacts of hydraulic fracturing	Environmental impacts associated with hydraulic fracturing could affect policymakers' decisions regarding natural gas production. The EPA is currently conducting a study to determine the extent to which damage has occurred and the future environmental risks involved with hydraulic fracturing. Severe impacts associated with drilling could lead to moratoriums, such as the one issued by the state of New York, or limits placed on drilling. These political actions would decrease the supply of available shale gas.
Technological advances in natural gas extraction	Over the past few decades, the oil and natural gas industries have been transformed into one of the most technologically advanced industries in the United States. With an annual technological growth rate of 2.4%, increases in efficiency rates for production and transmission, and availability factors for generating capacity can lead to higher overall generation and potential decreased prices.[13]
Natural disasters	Based on natural gas storage differentials, extreme weather implications will impact production and, thus, the price of natural gas. As a general rule, inventories above the seasonal norm decrease prices while inventories below the seasonal norm increase prices. As a result, any production disruptions will likely increase the price of natural gas. As extreme weather has been statistically above average in the past decade, technological advances have to some degree decreased the impact. Additionally, the forecasted shale supply and storage capacity have stabilized overall pricing.
Weather impacts	Temperatures statistically above or below average will impact futures prices of natural gas as well as demand based on necessary heating degree days (HDD) and cooling degree days (CDD). These prices are in large part contingent on NYMEX natural gas contract prices, which dictate actionable buy or sell signals in trading. Based on 30-year average data, a deviation of 8 degrees showed that changes in natural gas futures occur 89% of the time.[14]

Supply Side	Implications
Policy	As the challenges of energy efficiency and environmental implications of energy production permeate through the political and industrial arenas, the U.S. economy faces increased vulnerability to natural gas price fluctuations and the subsequent effect on demand. The current political arena is considering carbon taxes and impact fees that have the potential to slow shale production.

Demand Side	Implications
Exportation of natural gas	As of March 14, 2012, the U.S. Department of Energy (DOE) authorized Cheniere Energy's plan to export liquefied natural gas (LNG) from the Sabine Pass Terminal in Louisiana. The plan approved by the Department of Energy allows Cheniere Energy to export as much as 2.2 billion cubic feet of liquefied natural gas per day for the next 20 years.[15] The agency is also considering seven other applications for exporting natural gas, the approval of which would allow for exportation of the amount of gas equal to one-fifth of the United States' current consumption. According to the DOE, exportation could increase the price of natural gas by anywhere from 3% to 9% by 2035.[16]
Industrial uses	Natural gas is not only used as an energy source, but is also used in the production of products, including fertilizer, plastics, and pharmaceuticals. Many of these products would benefit from cheaper natural gas. A particularly large product market is plastic production. Ethylene feedstock used in plastic production is often dictated by the domestic supply of natural gas and crude oil.[17] A recent EIA report suggests that plastic production in the United States accounts for 331 million barrels of petroleum and 11 billion cubic feet of natural gas (1.5% of the U.S. natural gas consumption).[18] A by-product of shale formation, natural gas liquids (primarily ethane) can be used to create plastics.[19]

Demand Side	Implications
Coal-fired power plants	In 2011 and 2012, U.S. coal energy generation decreased by 9.3% as the number of coal-fired power plants has continued to shrink.[20] Energy suppliers cite increasing environmental regulation as the reason to shut down the power plants.[21] New environmental standards released by the EPA are set to reduce emissions from mercury and other toxic pollutions, while future regulations look to keep all power plant emissions equivalent to that of natural gas-fired plants.[22]
Compressed natural gas vehicles	As gasoline prices continue to rise, compressed natural gas (CNG) vehicles are emerging as a viable alternative to traditional transportation. CNG vehicles can be purchased new or converted from existing gasoline-powered vehicles. CNG vehicles provide cost savings from lower natural gas prices and are environmentally preferable to traditional gasoline.

Exhibit 9.3 Technological Advancements in Natural Gas Exploration and Extraction

Technological Advancement	Description	Benefit
Advanced 3D and 4D seismic imaging	This technology uses traditional seismic imaging techniques, combined with powerful computers and processors, to create a three-dimensional model of the subsurface layers.[23]	There is an accurate exploration and identification of natural gas reservoirs.
CO_2-sand fracturing	This technology is used to increase the flow rate of natural gas and oil from underground formations. CO_2-sand fracturing involves using a mixture of sand proppants and liquid CO_2 to fracture formations, creating and enlarging cracks through which oil and natural gas may flow more freely.	Because there are no other substances used in this type of fracturing, there are no "leftovers" from the fracturing process that must be removed. It does not damage the deposit, it generates no ground waste, and it protects ground water resources.[24]

Technological Advancement	Description	Benefit
Measurement While Drilling (MWD)	It allows the collection of data from the bottom of the well as it is being drilled.	This allows engineers and drilling teams access to up-to-the-second information on the exact nature of the rock, resulting in higher drilling efficiency and accuracy in the drilling process.
Pipeline construction	It employs sophisticated robotics devices that are propelled along the pipelines to evaluate the interior of the conducts. These "Smart Pigs" can test thickness, roundness, and signs of corrosion as well as detect minute leaks and any other defect along the pipeline.	This system secures natural gas operations and reduces maintenance and repair costs.

Exhibit 9.4 Summary of Possible Annual Impact of the Variables Affecting the Cost of Natural Gas Power Generation

Variable	Maximum Impact	Most Likely Impact	Minimum Impact
Availability factor	−0.5%	−2.4%	−3.5%
Variable operating and maintenance costs	14.5%	5.0%	−12.5%
Natural gas power price trend	5.5%	1.0%	−5.0%
Efficiency rate	0.0%	−4.0%	−10.0%
Carbon taxes	20.0%	17.0%	14.0%
Occurrence of carbon taxes	N/A	25.0%	N/A
Impact fees	30.0%	21.0%	16.0%
Occurrence of impact fees	N/A	10.0%	N/A

10

Slotting Pharmaceuticals in an Automated Dispensing Cabinet

Matthew J. Drake, Duquesne University

Mary Rucinsky is the hospital administrator for the Southeast Ohio Women's Hospital (SOWH) located in Damascus, Ohio. Part of her responsibility is ensuring that the hospital operates as efficiently as possible while maintaining the highest level of patient care. Based on some best practices shared at a recent hospital administrator conference, Mary decided that her hospital should possibly consider installing an automated dispensing cabinet (ADC) to improve its nurses' access to the most commonly prescribed medications in its postpartum recovery unit while reducing errors in administering the medication to patients. Several salespeople from companies producing ADCs have touted their potential cost savings in meetings with Mary, but she wants to conduct her own analysis to see how much her facility can realistically be expected to benefit from investing in this new piece of equipment.

Southeast Ohio Women's Hospital

SOWH is the premier facility for women's health services and care in the southeast Ohio area. SOWH is nationally ranked for the quality of its gynecological and obstetrics care. Approximately 5,000 babies are born at the facility each year, representing almost 70% of the total births in Columbiana County and the surrounding counties. The Department of Obstetrics has received major research grants from the National Institute of Health in recent years.

The postpartum recovery unit consists of 50 rooms located on the building's third floor. New mothers spend an average of two to three days in a recovery room before going home with their new babies. Mary thinks that this unit in particular can benefit from the installation of an ADC because its patients require a variety of medications throughout their stay and currently nurses must travel all the way down to the hospital's pharmacy to obtain each dose for their patients. The nurses can batch their trips to the pharmacy by collecting medications for several patients at once, but they have to be careful not to gather too many doses at the same time because this could lead to medication errors. A dedicated ADC installed in the postpartum unit on the third floor of the hospital should allow the nurses to retrieve medication for patients on an individual basis and should reduce the total time that each dose takes a nurse to collect.

Medication Profile

Patients in the postpartum recovery unit can require hundreds of different medications for their specialized needs as they recuperate from delivering a baby. Of course, some medications are prescribed more often than others. Mary has identified the 50 most commonly prescribed medications, which are listed in Exhibit 10.1, along with their individual dosage package dimensions. All of these medications fall into the following four package categories:

- **Small bags**—1" × 0.2" × 2.5"
- **Large bags**—2" × 0.2" × 3.75"
- **Small boxes**—2.75" × 1.25" × 5"
- **Large boxes**—3.5" × 1.5" × 6"

These 50 medications are the strongest candidates to be stored in the new ADC. None of the other medications stocked in the pharmacy are prescribed often enough to justify taking up the limited space in the ADC. These items would continue to be retrieved from the centralized hospital pharmacy even after the ADC is installed.

Strategy for Stocking the ADC

After talking with several vendor sales representatives and considering the specifications of various models, Mary has identified the Accu-Med ADC produced by Ascent Medical Solutions as the best fit for her facility. The storage space in the Accu-Med unit measures 24" × 48" × 18". The Accu-Med unit costs $40,000 to purchase, and it also carries a required maintenance contract of $5,000 per year that SOWH would have to fund. SOWH uses a 25% cost of capital to evaluate capital investments such as the ADC equipment. Based on the current rate of technological development in the medical equipment industry, Mary expects that the Accu-Med unit should last for five years before SOWH would need to upgrade the system to meet its current needs at that time.

A quick analysis showed that the time the postpartum recovery unit's nurses spend retrieving a dose of medication from the centralized hospital pharmacy costs $6 in labor cost. The monthly historical demand volumes for the 50 medications (provided in Exhibit 10.2) suggest that it costs SOWH approximately $25,000 per month to gather these items from the pharmacy. Mary estimates that the ADC will cut the cost of collecting a dose down to $1 due mainly to the nurses' reduction in travel time. Restocking a medication in the ADC should cost $30 because the nurse has to scan each dose individually both when it is taken from the pharmacy as well as when it is deposited into the ADC to ensure that no doses go missing. Some of these medications are controlled substances that are prime candidates for theft; as a result, it is imperative that SOWH maintains a high level of care when moving these items from one storage location in its facility to another.

Mary's first inclination was to use an equal space method to stock all 50 of the medications in the ADC. That is, she planned to allocate the same amount of storage space in the ADC to each item. While discussing this with one of the analysts who reports to her, the analyst suggested a different approach. He recalled a discussion in the logistics course that he took during his MBA program related to strategies for stocking a fast-pick area in the warehouse. According to the discussion, an equal space method (and an equal time method, for that matter) is suboptimal because it does not adequately balance the two main costs in order picking: (1) the cost of picking and (2) the cost of restocking. The analyst was also unsure that all 50 of the medications that Mary identified actually warranted occupying the precious space in the ADC. It was obvious to him that at least 40 medications should be slotted into the ADC, but perhaps a better solution would be to stock only 45 of the medications to allow for more doses of each one to be stored in the ADC, thereby reducing the need for restocking each item quite so often.

ADC Benefit Analysis

Mary was intrigued by the analyst's ideas; however, she was not entirely convinced that the proposed stocking strategy would actually be more beneficial than the simple equal space allocation method that she had planned to use. She asked the analyst to spend some time estimating the benefit that SOWH could realize by using each of the two proposed stocking methods to slot medication into the ADC. She also wanted to confirm that these savings would justify the required investment in the ADC equipment.

To complete the charge that Mary gave him, the analyst planned to complete the following tasks:

1. Determine the total monthly cost of picking and restocking all 50 medications in the ADC using the equal space slotting strategy.
2. Identify the optimal mixture of medications to stock in the ADC according to the following procedure, and determine the total monthly cost of picking and restocking these items. This procedure is proposed by Bartholdi and Hackman.[1]
   a. Compute the labor efficiency for each medication using the following formula:

   $$\text{Labor Efficiency for Item } i = \frac{\text{Number of Picks of Item } i \text{ per Month}}{\sqrt{\text{Volume of Demand for Item } i \text{ per Month}}}$$

   b. Rank the items in order of decreasing labor efficiency. Labor efficiency represents the claim that each item has to be stored in the fast-pick area of a warehouse. An item with a higher labor efficiency generates more benefit from being stocked in the fast-pick area instead of a bulk storage area.
   c. Consider the first 40 items with the highest labor efficiency. This step will determine the optimal quantity of each of these items to put into the ADC as well as the total cost of this allocation.
      i. Determine the recommended volume of each of the 40 items to stock in the ADC (v_i) using the following equation, where V is the total volume of storage space available in the ADC.

      $$\text{Recommended ADC Volume for Item } i \ (v_i) = \left(\frac{\sqrt{\text{Volume of Demand for Item } i \text{ per Month}}}{\sum_{j=1}^{40} \sqrt{\text{Volume of Demand for Item } j \text{ per Month}}} \right) * V$$

      ii. Compute the total monthly picking and restocking cost generated by stocking the volumes of the 40 items in the ADC that were determined in the previous step. Each of these monthly costs can be computed with the following formulas:[2]

Monthly Picking Cost for Item i in the ADC =
Cost per Pick from ADC × Monthly Picks for Item i

Monthly Picking Cost for Item j from the Pharmacy =
Cost per Pick from Pharmacy × Monthly Picks for Item j

Monthly Cost to Restock Item i in the ADC =

$$\text{Cost per Restock} * \frac{\text{Volume of Demand for Item } i \text{ per Month}}{v_i}$$

    d. Repeat step C for the scenario where the top 41 medications ranked by labor efficiency are stocked in the ADC. Continue repeating this step until all 50 medications are slotted into the ADC. The recommended slotting strategy is the plan with the number of items that minimizes the total monthly picking and restocking cost.

3. Determine the estimated monthly savings generated from the equal space allocation (from step 1) and the analyst's slotting procedure (from step 2) compared with the current cost of collecting all medications from the pharmacy without installing the ADC.

4. Estimate the net present value of SOWH investing in the ADC under each of the two allocation strategies. Determine if SOWH should invest in the ADC; if so, recommend the allocation strategy that should be used.

Endnotes

1. J. J. Bartholdi and S. T. Hackman (2014). *Warehouse & Distribution Science*, version 0.96. Available online at www.warehouse-science.com.

2. Note that if 40 items are slotted into the ADC, the remaining 10 items from the list must be retrieved from the centralized hospital pharmacy whenever they are prescribed for a patient.

Exhibits

Exhibit 10.1 Dimensions of the 50 Medications That Are Candidates for Stocking in the ADC

SKU	Length (in Inches)	Width (in Inches)	Height (in Inches)
1	1	0.2	2.5
2	1	0.2	2.5
3	1	0.2	2.5
4	1	0.2	2.5
5	1	0.2	2.5
6	1	0.2	2.5
7	1	0.2	2.5
8	1	0.2	2.5
9	1	0.2	2.5
10	1	0.2	2.5
11	1	0.2	2.5
12	1	0.2	2.5
13	1	0.2	2.5
14	1	0.2	2.5
15	1	0.2	2.5
16	1	0.2	2.5
17	1	0.2	2.5
18	1	0.2	2.5
19	1	0.2	2.5
20	1	0.2	2.5
21	2.75	1.25	5
22	2.75	1.25	5
23	2.75	1.25	5
24	2.75	1.25	5
25	2.75	1.25	5
26	2	0.2	3.75
27	2	0.2	3.75
28	2	0.2	3.75
29	2	0.2	3.75
30	2	0.2	3.75
31	2	0.2	3.75

SKU	Length (in Inches)	Width (in Inches)	Height (in Inches)
32	2	0.2	3.75
33	2	0.2	3.75
34	2	0.2	3.75
35	2	0.2	3.75
36	2	0.2	3.75
37	2	0.2	3.75
38	2	0.2	3.75
39	2	0.2	3.75
40	2	0.2	3.75
41	2	0.2	3.75
42	2	0.2	3.75
43	2	0.2	3.75
44	2	0.2	3.75
45	3.5	1.5	6
46	3.5	1.5	6
47	3.5	1.5	6
48	3.5	1.5	6
49	3.5	1.5	6
50	3.5	1.5	6

Exhibit 10.2 Monthly Picks and Units Demanded for Each of the 50 Medications That Are Candidates for Stocking in the ADC

SKU	Picks per Month	Demand per Month
1	65	68
2	67	73
3	166	166
4	2,019	3,564
5	175	215
6	66	94
7	15	30
8	99	99
9	469	931
10	413	413
11	1,041	1,921

SKU	Picks per Month	Demand per Month
12	493	887
13	241	290
14	456	708
15	258	267
16	574	628
17	71	90
18	251	264
19	59	108
20	362	391
21	143	512
22	201	307
23	385	394
24	1,203	1,359
25	13	39
26	498	561
27	319	333
28	452	475
29	57	77
30	783	851
31	20	65
32	11	39
33	805	805
34	571	571
35	21	26
36	280	344
37	301	319
38	98	98
39	306	324
40	236	254
41	323	332
42	592	997
43	521	932
44	297	310
45	93	202
46	23	37

SKU	Picks per Month	Demand per Month
47	258	473
48	302	429
49	807	923
50	696	927

Part III
Analytics in Accounting and Finance

11 Tax Ramifications of S Corporation Shareholder Termination or Change of Ownership Interest 123

12 Charleston Rigging 131

13 Equipment Purchase and Replacement Strategy at the Fayette China Company 139

11

Tax Ramifications of S Corporation Shareholder Termination or Change of Ownership Interest

Elizabeth Conner, University of Colorado – Denver

Doug Laufer, Metropolitan State University of Denver

Introduction

Under U.S. tax law, there are four choices for the type of business entity: sole proprietorship, partnership, limited liability company, and corporation. Within the fourth choice (corporation), there is an additional option, a C corporation or S corporation. The names denote subchapters of the Internal Revenue Code that contain the "rules" for each type of corporation.

C corporations include all publicly traded corporations and many privately held companies as well. A C corporation can have an unlimited number of shareholders, has limited liability, and is taxed on its income. Additionally, employees are taxed on salaries paid by the C corporation and shareholders are taxed on dividend distributions. Because this "double taxation" was viewed as unfair to family-run businesses and other small business corporations, Congress created the Small Business Corporation Investment Act in 1958. The rules were codified for the small business corporation in Subchapter S of the Internal Revenue Code.

A Subchapter S corporation (referred to as an S corporation) is a hybrid type entity because it has characteristics of both a regular C corporation and a partnership. Many small businesses today are structured as S corporations because an S corporation affords the same legal protection as a regular C corporation and is considered a pass-through entity similar to a partnership where the profits/losses are passed directly to the shareholders. This allows an S corporation to avoid double taxation on the corporate income. With these benefits come limitations on S corporation formation and ownership rules. There are also unique issues S corporations face in the allocation of profit/loss between shareholders.

As a pass-through entity, an S corporation must allocate its profit/loss for the year pro rata (also known as daily proration) to its shareholders based on the number of outstanding shares each shareholder owns during the tax year.

If a shareholder sells his or her shares during the year, the shareholder will use the same daily proration allocation to report his or her share of the S corporation profit/loss for the days the shareholder owned the stock, including the day of sale. In addition, the shareholder will report a capital gain or loss from the sale of the shares in the S corporation.

The IRS allows another method of allocating profit/loss of an S corporation when a shareholder sells his or her shares in the corporation during the year. This is known as the "closing of the books method." This method treats the taxable year of the S corporation as if it consists of separate taxable years. The closing of the books method is only permitted when elected by all affected shareholders (those with changing ownership during the year).

For a shareholder who terminates his or her ownership interest in the S corporation during the year, the S corporation makes the election by attaching a statement to the tax return stating that it is electing for the taxable year under Treas. Reg. § 1.1377-1 to treat the taxable year of the S corporation as two taxable years for the shareholder who terminates his or her entire interest in the company during the year.

For a shareholder who is selling a partial interest, the disposition must be deemed a qualifying disposition where the shareholder sells 20% or more of his or her ownership interest within a 30-day period. Upon agreement by all affected shareholders, the S corporation can make an election by attaching a statement to the tax return stating that it is electing for the taxable year under Treas. Reg. § 1.1368-1(g)(2)(i) to treat the taxable year as if it consists of two separate taxable years.

The S corporation and its shareholders should be aware that the daily proration method used in allocating the profits/losses of the S corporation each year to its shareholders is not necessarily the only method allowed in a year that a shareholder sells his or her interest. The alternative method known as "closing of the books" may give a much better income tax result to the affected shareholders over the daily proration method. Tax planning is essential in this area because the amount of income taxes levied on the affected shareholders can be vastly different.

Decision Scenarios and Discussion

This section presents a case about a privately held company and its shareholders. It starts with a description of the company and its recent history. Then it discusses two decision scenarios to examine the tax implications of a change in a shareholder's interest in the company. The facts used in this case are a fictionalized simplification of an actual tax matter researched by a practicing CPA for his clients.

Background Information

Mary Simpson and Jack Depew met a number of years ago at an estate sale auction. Mary and Jack shared a passion for antique furniture. They each enjoyed the process of hunting for antique furniture as well as restoring and selling pieces they had acquired. They each had acquired quite a collection of items. They began combining their efforts to buy, restore, and sell items and after a number of months decided to formally go into business together.

Mary and Jack, who are unrelated taxpayers, began operating an antique furniture company called Antique Haven, Inc., in 2007. After consulting with a CPA, Mary and Jack decided to incorporate their business and filed the necessary election to be treated for tax purposes as an S corporation with a calendar year-end. The company issued an equal number of common shares to Mary and to Jack in exchange for $6,000 invested by each of them.

During the period from 2007 through 2013, Antique Haven's business grew and prospered. Mary and Jack reported their share of the profits of the company on their personal tax returns in accordance with the S corporation rules pertaining to daily proration. Each year, the company paid "dividend" distributions to Mary and Jack. These distributions were reported as reductions in Mary and Jack's stock basis in the company and, therefore, were not taxable to them. After 2012, a downturn in the economy began to affect Antique Haven's business. The company struggled and reported losses in both 2013 and 2014. In addition, due to cash flow constraints, the company suspended paying distributions to Mary and Jack.

At the beginning of 2015, Mary and Jack each had a basis in the stock of Antique Haven, Inc., of $200,000. During the first five months of 2015 (January 1 to May 31, 2015), the company reported a loss (non-separately computed) of $319,969.

In the summer of 2015, Antique Haven decided to engage in an active advertising campaign to try and improve its operating performance. The advertising proved very successful and with the bustling holiday shopping season, Antique Haven was able to negate the loss from earlier in the year and then some. The company was happy to report an overall profit (non-separately computed) of $470,120 for the year 2015.

Decision Scenario One: S Corporation Termination of a Shareholder's Interest

Mary had become more and more fed up with Antique Haven's poor performance over the past few years as the losses continued to mount each month. She reached her breaking point in May of 2015 and sold her entire interest of 300 shares to Sally Benson on May 31 for $100,000.

It is now the end of 2015, and Antique Haven is required to prepare its tax return and Form 1120S. The company must also report the income and expenses to its current shareholders and former shareholder, Mary.

Recall that a pass-through entity reports its profit/loss to the shareholders using a pro rata (daily proration) method of allocation. When a shareholder terminates his or her interest during the year, daily proration is required unless a special election is made to use the closing of the books method.

Required Analysis and Discussion

1. Describe the pro rata or daily proration allocation method and the closing of the books method when a shareholder terminates his or her interest in an S corporation during the year. As part of the closing of the books method, describe the special election requirement.
2. Using the daily proration method for 2015, determine the tax impact of this transaction on Mary, Jack, and Sally.

3. Using the closing of the books method for 2015, determine the tax impact of this transaction on Mary, Jack, and Sally.

Decision Scenario Two: S Corporation Change in a Shareholder's Interest through Sale of Stock Other Than Termination of an Interest

Similar rules apply to an S corporation's shareholders when a shareholder sells a portion of his or her shares in an S corporation; that is, a pro rata allocation (daily proration) is required. However, if the selling shareholder meets the qualifying disposition rules set forth in Treas. Reg. § 1.1368-1(g)(2)(i) and all affected shareholders consent, the S corporation can elect under Treas. Reg. § 1.1368-1(g)(2)(iii) to use the alternative allocation method of closing of the books.

Refer to the details of the original scenario. Assume now that Mary is still dissatisfied with the company's performance but not so much that she wants to liquidate her entire interest in the company. She decides to sell 180 of her 300 shares to Sally on May 31, 2015, for $60,000. Mary downsizes from owning 50% to owning 20% of Antique Haven, Inc., and Sally becomes a 30% owner of Antique Haven, Inc. Jack remains a 50% owner.

It is now the end of 2015, and Antique Haven is required to prepare its tax return and Form 1120S. The company must also report the income and expenses to its current shareholders, including Mary.

Required Analysis and Discussion

1. Describe what is meant by a "qualifying disposition." Is Mary's disposition of her 30% interest in the stock of Antique Haven considered a qualifying disposition?
2. Using the daily proration method for 2015, determine the tax impact of this transaction on Mary, Jack, and Sally.

3. Using the closing of the books method for 2015, determine the tax impact on Mary, Jack, and Sally. As part of your answer, describe the special election that must be made by Antique Haven in order to use the closing of the books method in this case.

Conclusion

This case provides a vehicle for performing tax research of the primary tax authorities related to a situation where a shareholder of an S corporation either terminates or changes his or her ownership interest. Recognizing how to apply underlying tax authority to a specific set of facts, synthesizing the research findings and calculations, and providing written communication to explain and support the tax impact under the given case scenarios are the important components of this case study.

References

Internal Revenue Code § 302

Internal Revenue Code § 303

Internal Revenue Code § 1366

Internal Revenue Code § 1377

IRC Treasury Regulations § 1.1368-1

IRC Treasury Regulations § 1.1377-1

12

Charleston Rigging[1]

Thomas McCue, Duquesne University

In early January 2014, Robert Keyes, the owner of Charleston Rigging, began to think about expanding his business. Charleston Rigging was a small, privately held company located in Charleston, South Carolina. It had two main divisions. One manufactured and sold chains, wire rope, and frames used to unload and load containers in the maritime industry. The other division was a metal fabricating shop where they did metal work, forging, and other kinds of custom metal work on a job-by-job basis. Several years ago, Robert had purchased an old warehouse adjacent to his present site to be used as a storage facility for finished goods inventory. Over the past five years, Charleston Rigging began to build a network of marine distributors who now carried their inventory. Thus by 2014, it was clear that using the warehouse to store inventory made little sense. Robert began to think about using the space to expand his business. Both product lines were running close to full capacity, although he actually believed that he was losing sales in the job shop due to its lack of capacity.

History of Charleston Rigging

Robert Keyes's great-grandfather, James Keyes, founded Charleston Rigging in 1909, shortly after President Theodore Roosevelt identified a site on the Cooper River as a good spot for a navy yard. James Keyes was only 20 years old when he began to supply goods for U.S.

Navy ships that called the Charleston Navy Yard home. Initially, the firm supplied food, including fresh vegetables, paper supplies, and small tools. During World War I, James Keyes purchased a small forge and metal shop just north of the naval base. There he began to make anchor chains, rigging, and pulleys in addition to fabricating small parts. As the shipbuilding and repair facilities were expanded, the company's sales grew accordingly. Prior to World War II, the company began to concentrate more on anchor chains, chain slings, marine rope, and rope bumpers, although they still did a great deal of work fabricating small parts for the shipyard. World War II saw an expansion of the company's sales as the yard built many ships and converted civilian ships to military uses. After World War II, Robert's grandfather, John Keyes, purchased an adjacent property and built a new shop. The original shop was then upgraded and used to make parts used in making chain slings and anchor chains. John Keyes pared the product line somewhat, focusing mainly on anchor chains and chain slings for loading and unloading ships, although the firm still fabricated small jobs for the navy yard.

The firm's first big challenge came in the early 1960s. The navy shifted shipbuilding to other yards and Charleston Navy Yard became a facility to repair nuclear submarines. A submarine did not use anywhere near the number of products from Charleston Rigging that a naval surface ship did. True, they still machined some small parts for the submarines, but the technical sophistication of the metals used in nuclear submarines limited the parts fabrication market. Although the business was still profitable, John Keyes and his son, Andrew, began to market chain slings to merchant ships through marine supply firms in New Orleans, Savannah, and later in Norfolk and Jacksonville. This expansion forced the firm for the first time to think about inventory management because up until then they really were a job shop only manufacturing an item when they had an order. Sales recovered but were still below the level they reached during the war.

Andrew and his son, Robert, then faced their next big challenge in the early nineties when the navy announced that they were closing the shipyard. This decision effectively ended the forging and machining of parts for the navy. When they heard the news, they thought the impact would be devastating. This supposed misfortune, however, turned out to be a blessing in disguise. The state of South Carolina took over the navy base and expanded the port facilities, allowing it to handle more container ships. In response to the closing, Charleston Rigging shifted focus to the merchant marine industry. They expanded their chain and rigging business. The job shop work fell off but recovered nicely by 2012. Sales reached a new high in 2013. The port of Charleston had become the second busiest container port on the East Coast and the sixth busiest port in the country.

A strong emphasis on customer service no doubt played a role in the firm's success. Robert, in particular, believed service was part of his product. He frequently carried slings down to the port for a customer who had a sling break. He also spent time talking to trucking companies, stevedores, and ship captains about his products and what he could do to improve them. He loved to tinker with new ideas and designs. He opened a sales office in Georgetown, South Carolina, a port about 60 miles north of Charleston that had a large paper mill and a steel mill. He was working on chain straps for trucks that delivered logs to the paper mill in Georgetown. He worked hard to cultivate a good relationship with his distributors.

In 2009, when Robert Keyes was 40 years old, he took over the business when his father died suddenly. At the time of his father's death, Robert had spent 18 years working in the business. He began with the firm after he had graduated from college. His father had insisted that Robert rotate through every job in the plant from operating machines that made chain slings to running the firm's bookkeeping and computer system. His father had believed in very conservative financial policies. He did not like to carry any debt. Whenever the company had expanded, cash was used to finance the expansion.

Although not as tradition bound as his father, Robert did not like debt either. He had a line of credit with a local bank, but to date he had never used it.

The Opportunity

Robert thought that the space that was freed up could be used to increase sales. Production of chains, slings, and cargo nets offered an opportunity because he thought his firm's reputation for quality attracted sales and would continue to do so if only from his existing customer base. Alternatively, the space could be used to increase sales in metal fabrication. He had no idea exactly how much new business this would generate. He guessed that about one quarter of his customers in metal were turned away because he did not have the capacity or because the job would take too long to complete.

Nationally, Charleston Rigging had never had a large market share in the chain and sling market. Robert deliberately attempted to operate "under the radar screen" of several large national competitors. He charged a premium for his slings, but many shipping firms and stevedoring companies didn't mind paying more because Charleston Rigging had a reputation for quality, service, and ease of use. Like his father, Robert did not spend much on advertising nor would he be expected to begin now. He did take out ads in a few marine industry trade publications; however, most of his sales were to repeat customers. When a new customer called to open an account, Robert usually asked how the customer heard about their products. The most common answer was through word of mouth.

The Marketing Study

In the spring of 2013, Robert had ordered a marketing study that focused mainly on the chain and sling division. The key points of this study were:

1. The chain and sling business was forecasted to have 20% growth with a 20% chance, 5% growth with a 40% chance, and 40% chance of no growth over its 2013 base level of $3,380,104.
2. The chain and sling division's cost of materials has traditionally run 68.8% of sales, while labor cost has run 7.4% of sales. For the metal fabrication business, labor has run 18% while material was 32% of sales.
3. The study felt that forecasting sales for the metal fabricating division was extremely difficult. They believed that sales would increase, but had no real data to make a good forecast. (Robert guessed it could rise 15% above its 2013 number of $3,120,096.)
4. The study also cautioned against putting too much faith in the forecast for slings and chains. Container ships were expanding in size, which should be an indicator of solid growth; however, many believed larger container ships would actually cause more congestion in ports, forcing many customers to look for other methods to ship products.
5. An increase in labor would add $5,000 to the fixed costs attributable to the metal fabricating business.

Charleston Rigging had no formal financial officer. The firm had a clerk who made the journal entries, managed accounts receivable, and completed payroll. Robert relied heavily on a local CPA firm, Davenport Associates, to manage the firm's finances. Ken Lewis, a partner with Davenport, prepared quarterly and annual GAAP financials. He also worked with Robert on preparing annual budgets, inventory schedules, and other internal financial statements. Davenport had recently completed the 2013 financials. (The firm's fiscal year ended in September.) Sales for 2013 were more than $6.5 million, up substantially over 2012 levels. (See Exhibit 12.1 for 2013 financial statements.) Since 2008, sales had positive growth in every year. Lately, the metal fabricating business had begun to have flat growth or very slow growth. Robert blamed this slow growth on the fact that the metal fabrication business was operating at full capacity. He was convinced that if he added capacity in this line he would be able to turn the new capacity into more sales. Meanwhile, chain sling sales had actually contributed more to growth over the period. Although chain slings were basically a commodity product, he achieved most of that growth by adding new distributors (see Exhibit 12.2). Sales by distributors were rising faster than any other geographic channel.

Earlier in the week, he had called Ken Lewis, his CPA, to ask him what he thought. Ken had told him that he should gather as much information as possible about the proposal: costs, capital employed, labor hours required, inventory levels, and space needs. Once the data was available, Ken would put together a capital budgeting analysis of the project. He reminded Robert that these two options for using the open space were mutually exclusive. Ken also told Robert to analyze the decision using a contribution analysis.

Robert now had begun to work on the project.

Endnote

1. The purpose of this case is to illustrate a management decision for class analysis and discussion. The persons and company in this case are fictitious.

Exhibits

Exhibit 12.1 Charleston Rigging Financial Statements Dated September 30, 2013

Income Statement	
Sales	$6,500,200
Variable costs	
Material costs	3,775,872
Labor costs	624,124
Contribution margin	2,100,204
Fixed expenses	305,090
Net operating income	1,795,114
Tax	183,974
Net income	$1,611,140
Balance sheet	
Assets	
Cash	$63,238
Accounts receivable	566,030
Inventories	942,529
Total current assets	1,571,797
Net fixed assets	2,003,022
Total assets	$3,574,819
Liabilities and equity	
Payables	$390,012
Accruals	195,006
Total current liabilities	585,018
Equity	2,989,801
Total claims	3,574,819

Exhibit 12.2 Charleston Rigging Geographic Sales Distribution Summary

Geographic Sales Distribution	Charleston	Georgetown	Marine Distributors
Sales	33%	12%	55%
Slings	34%	6%	60%
Metal shop	97%	3%	0%

13

Equipment Purchase and Replacement Strategy at the Fayette China Company

Matthew J. Drake, Duquesne University

Robert Ross is the plant manager at the Fayette China Company's facility in Charleroi, Pennsylvania. Among his many responsibilities is the decision on when and how to replace the capital equipment used to manufacture the company's iconic Shangri La line of dinnerware products. Much of the equipment in the facility is used heavily, and as a result, it must be replaced at a minimum approximately every ten years. Sometimes, however, it is replaced far more frequently than that if the costs of doing so can be justified.

One of the kilns in the facility has just come up for replacement, and the Fayette China Company's executive board has mandated that all new kilns installed in the firm's facilities must be state-of-the-art computerized kilns. These computerized kilns allow workers to control the heating of the kiln more precisely, which enables the facility to produce higher-quality output with fewer defects. Robert has been talking with sales representatives from two different suppliers to try to determine which of their models would be the best choice for his facility.

Company Background

The Fayette China Company was founded by brothers Carl and Bartholomew MacArthur in Monessen, Pennsylvania, in 1876. The MacArthur brothers had settled in Monessen, a small town on the Monongahela River approximately 27 miles southeast of Pittsburgh, after emigrating from the Scottish Highlands five years earlier. Their company started as a distributor of pottery, both made in nearby factories as well as a popular white variety imported from England.

After a few years of distributing pottery for other producers, the brothers decided to open their own manufacturing plant. They purchased a plot of land on the Monongahela River in Monessen for $400 in 1878 and opened their new plant with three kilns in July of 1879. The company grew quickly and its products came to be known for their quality and attractive design.

The company continued to grow over the next few decades. The MacArthur brothers expanded their operation at the Monessen plant by adding kilns every two or three years, but by the turn of the century, it was clear that they were running out of space in the existing facility. The plot of land was too small to expand the building on the current site, so the owners looked for nearby tracts of land that would accommodate the company's current as well as its future space needs. They found what they were looking for across the Monongahela River in Charleroi, and they opened their new facility there in 1909 with 128 kilns. This is the same site where the current facility still stands, although the original building has subsequently been expanded several times to attain its current footprint. The company has also built several adjacent buildings on the site to house corporate and administrative business functions.

Equipment Replacement Strategy

Although Robert did have to determine the model of computerized kiln to install, the decision was actually a little more difficult than choosing from the two options. Each kiln would have its own maintenance and replacement schedule based on its various cost parameters and the trajectory of its loss of value over the years. Robert wanted to first determine the total cost of operating and possibly replacing each of the kiln models over a ten-year planning horizon. The Fayette China Company uses a discount rate of 25% in all of its capital budgeting decisions. Then based on the total cost of each model's operation and replacement strategy, he could determine the best choice for his facility. For this analysis, he decided to assume that if a particular kiln were going to be replaced, it would be replaced by the same model—although this was not necessarily a restriction that he was required to follow in practice. For an ease of comparison, Robert assumes that the company will purchase a new kiln at the end of year ten, regardless of the age of the existing kiln.

Robert knew that his decision hinged on the relative magnitude of the cost and operating parameters of the two models. (We call them Model A and Model B for simplicity.) These specifications are summarized in the following sections.

Initial Cost

The two kilns under consideration have similar initial costs, but small differences can add up if the equipment were to be replaced several times over the planning horizon. Model A costs $75,000, and Model B costs $83,000. Both of these costs include charges for shipping to the Charleroi facility and for installation. Based on Robert's experience in the industry, he expects the cost of both of these models to increase by 5% per year over the next ten years.

Operating and Maintenance Cost

The yearly operating cost of a kiln is dependent upon the total amount of hours that the kiln is running per year. The Fayette China Company uses a capacity lag strategy whereby its kilns are only installed if 100% of their effective capacity is required. Therefore, Robert assumes that the kilns will be run at full capacity each year. The operating cost for Model A is $8,000 per year, and the cost for Model B is $7,500 per year.

In addition to operating the kiln, the plant must perform periodic maintenance and repair on its kilns. This maintenance and repair gets more expensive the older the kiln is. Yearly maintenance and repair costs for each of the kiln models at various ages are provided in Exhibit 13.1.

Salvage Value for Used Machinery

If Robert decides to replace a kiln before the end of its useful life, the company can sell the used machine on the secondary market. Robert estimates that Model A will lose 20% of its current value for each year of use, and Model B will lose 30% of its current value each year. These annual decreases in salvage value are summarized in Exhibit 13.2 as percentages of the kiln's original purchase price.

Equipment Purchase Decision

Armed with the specifications of each kiln supplier's quote, Robert was ready to analyze the two quotes to determine which model of kiln would work better for his plant. As he started the analysis, though, he began to question some of the assumptions that he was making. They certainly seemed reasonable, but what if the future unfolded somehow differently than he expected? Ten years was a long time.

What if the prices of the new machines changed at a different rate that was not constant year after year? What if maintenance and repair costs were larger or smaller than he thought? What if the value of old machines on the secondary market differed from Robert's estimates?

Although he planned to conduct his analysis with the estimated parameters he had already established, Robert wanted to investigate if and how his decision would change if there were some uncertainty in his estimates. By taking into account this uncertainty, he would feel more confident that the decision he had made had considered many different future scenarios.

Exhibits

Exhibit 13.1 Yearly Maintenance and Repair Costs of Each Kiln Model Depending on the Age of the Kiln

Age of Kiln in Years	Model A	Model B
1	2,000	1,500
2	3,000	2,250
3	4,500	3,375
4	6,750	5,063
5	10,125	7,594
6	15,188	11,391
7	22,781	17,086
8	34,172	25,629
9	51,258	38,443
10	76,887	57,665

Exhibit 13.2 Remaining Salvage Value as a Percentage of the Original Value

Age of Kiln in Years	Model A	Model B
1	0.8000	0.7000
2	0.6400	0.4900
3	0.5120	0.3430
4	0.4096	0.2401
5	0.3277	0.1681
6	0.2621	0.1176
7	0.2097	0.0824
8	0.1678	0.0576
9	0.1342	0.0404
10	0.1074	0.0282

Part IV
Analytics in the Public Sector

14 Using Regression to Improve Parole Board Decisions 147

15 Redesigning Pittsburgh Port Authority's Bus Transit System 153

14

Using Regression to Improve Parole Board Decisions

Wendy Swenson Roth, Georgia State University

Collin, the recently appointed chief information officer for the Colorado Department of Corrections, was just returning from a meeting with Sadie, the director of the Colorado Department of Public Safety. Sadie requested the meeting with Collin. Sadie was concerned about the performance of the board of paroles on two fronts. First, new government regulations about to take effect required an increased amount of documentation detailing the board's performance. This would increase the workload and lead to greater scrutiny of the board's decisions. Second, state budget cuts required the board to make decisions even quicker than before.

In their meeting, Sadie explained that it seemed every article and journal crossing her desk or into her email these days discussed how analytics was being used to make better decisions. Her knowledge about analytics was limited, but she knew data availability was required for analytics. She reasoned they had lots of data about the successes and failures of previous parole board decisions and that analytics should be able to help them make better decisions in the future. "This is where you come in, Collin," she said. "I would like you to investigate and propose how we can use analytics to help our parole board improve their decision-making process."

When Collin returned to his office, he sent a message to two of his analysts, Tom and Jerry, scheduling a meeting to get going on the new project. In the meantime, knowing that projects like this can go off in the wrong direction if they are not clearly defined and tracked, he noted the phases of problem solving from a favorite textbook (Exhibit 14.1) on his board to prepare for the meeting with Tom and Jerry.

Collin thought Sadie had correctly recognized the problem. He would discuss the problem with Tom and Jerry and then task them with coming up with a clear problem definition. Next, their job would be to structure the problem and report back to him.

The next day, after meeting with Collin, Tom and Jerry found an empty conference room to brainstorm the project. As a first pass at defining the problem, they agreed on "an accurate and efficient way to predict whether a person should be granted parole or not." This would be based on determining the likelihood of a person committing a future crime.

They discussed that multiple methods could be used to model this problem. Because they were faced with classifying people based on many independent variables, some form of regression seemed the perfect fit for this problem. Both of them decided to spend time thinking about possible dependent and independent variables. They would meet the following day to discuss their ideas.

Assuming they weren't the only parole organization looking to use analytics to address this issue, Tom decided to do some research. He located an article in *The Wall Street Journal* discussing the use of software applications to help guide the decision-making process on parole decisions. This article reinforced that they were in-line with the direction other parole boards were taking, evaluating modern risk assessment models with the goals of improving decisions and reducing costs.

Reading the article reinforced that improved decisions meant two things, not granting parole to those who will re-offend and also granting parole to those who won't re-offend. Getting this decision correct saves money by decreasing the workload for law enforcement officers and also prison costs by granting parole to those who won't re-offend.

Some of the issues in the article that Tom felt were critical to address include the following:

1. The impact on the parole decision process

   Previously, decisions were based on the expertise, experience, and intuition of the parole board members. The software applications provide an additional tool, increasing the amount of information available for each decision. Patterns, based on historical data, are used to predict the likelihood of re-offending based on an individual's characteristics. Research by criminologists is also used in model development.

   The availability of this information impacts the decision process. Situations will arise where the results of the model contradict the instincts of the parole board members. Support, training, and a decision-making process that helps to address these contradictions are critical for a successful implementation.

2. Understanding and explaining the difference between determining appropriate punishment and predicting the likelihood of re-offending

   "Some people are surprised to learn that offenders who we think of as the worst offenders—murderers and sex offenders—have some of the lowest recidivism rates," said Lee Scale, former director of internal oversight and research at the California Department of Corrections and Rehabilitation. This line caused Tom to stop and think again about what problem they were trying to solve. Determining whether a person will re-offend is different than punishment based on the severity of the crime. Implementation of the model may result in decisions that would be viewed critically by victims of violent crimes, their families, and the public in general. Again, these are issues that will need to be addressed for a successful implementation.

Despite these issues, Tom felt confident that this project would have a positive impact. The article included statistics from some states using computerized assessments that saw reductions in prison populations and recidivisms in multi-year studies. He felt this information would help gain acceptance for implementing this model.

This gave him a lot to talk to Jerry about at their next meeting.

Study Questions

Your group is responsible for performing Tom and Jerry's assigned tasks.

To provide more research on the problem, you can read *The Wall Street Journal* article:

> Joseph Walker, "State Parole Boards Use Software to Decide Which Inmates to Release: Programs Look at Prisoners' Biographies for Patterns That Predict Future Crimes," *The Wall Street Journal*, October 11, 2013, http://online.wsj.com/news/articles/SB10001424052702304626104579121251595240852.

Once you feel you understand the problem and possible ways to address it, answer the following questions:

1. How would you define the problem?
2. When structuring the problem, discuss the variables (independent and dependent) for a regression model that could be used by a parole board to determine risk of re-offending.
    a. Determine possible dependent variables.
    b. Determine ten possible independent variables, including sources required to obtain this information.
3. As an example, explain how the characteristics of one hypothetical individual could be calculated in your regression equation.

4. Discuss how the results from the regression model can be interpreted and used in making a decision about whether to grant parole. What specific issues need to be addressed about using a regression model in this situation?

5. Discuss other types of models that could be used to address this problem.

6. What specific ethical issues need to be addressed when implementing a regression model in this situation?

Closing Comments

As highlighted in *The Wall Street Journal* article and the preceding questions, the decision-making process is more complex than just developing the model. All the steps must be addressed to increase the likelihood of a successful implementation. Analytic models are meant as tools to improve the decision-making process, not to replace our role in the process.

Exhibit

Exhibit 14.1 Steps of a Decision-Making Process

1. Recognizing a problem
2. Defining the problem
3. Structuring the problem
4. Analyzing the problem
5. Interpreting results and making a decision
6. Implementing the solution

Data source: James R. Evans, *Business Analytics, Methods, Models and Decisions*. (Upper Saddle River, NJ: Pearson, 2013, pp. 22–24).

15

Redesigning Pittsburgh Port Authority's Bus Transit System

Ersin Körpeoğlu and Fatma Kılınç-Karzan, Carnegie Mellon University

Introduction

Allegheny County Port Authority is the main organization that oversees public transportation in Allegheny County and the city of Pittsburgh. The Port Authority's 2,400 employees operate, maintain, and support bus, light rail, incline, and paratransit services for nearly 230,000 daily riders. It is the second largest such system in Pennsylvania and the eleventh largest in the United States.

Due to a statewide transportation funding crisis, the Port Authority is facing the largest budget cut in the agency's 48-year history. If no solution is found, 46 venerable bus routes serving several of its townships, including Mt. Lebanon, Coraopolis, Green Tree, Mt. Washington, Oakmont, Edgewood, and Sewickley, will be eliminated. Riders who aren't stranded will pay more—the authority plans a 25- to 50-cent increase in different zones. According to a summary released by the Port Authority, as part of a 35% reduction in service hours, all of the authority's current 102 bus and rail routes would be scaled down, some ending altogether and others encountering deep drops in off-peak and weekend services. In the current version of the bus system, most buses serve between a point of origin in downtown Pittsburgh and a focal point, which is a highly demanded point for transportation, and pass through many townships in the meantime. For example, bus 61A serves Swissvale, Wilkinsburg, and also several demanded sights of Pittsburgh, such as Oakland and Downtown. The downside of such a system is that a bus tour takes approximately 2 hours even though most of the passenger traffic is between focal points. Therefore, less-demanded regions are either served infrequently or many redundant buses are assigned to achieve a certain service rate to those less-demanded regions.

To retain service, albeit at a minimal level, at all townships, and at the same time eliminate redundant bus routes, the Port Authority has started a new "park and ride" project.[1] As the name of the project implies, there will be main parking facilities called "hubs" in some of these townships and passengers will park their own cars in one of these hubs, and then use the bus for the remainder of their trip. Additionally, to serve customers with no cars, the Port Authority will assign shuttles that will carry passengers to parking hubs. For example, if there is a parking facility in Ross, a passenger who lives in Ross and wants to travel to Downtown will first get to the parking facility via either a shuttle or a car, and then use an express bus between the parking facility and Downtown. This way, the Port Authority plans to eliminate most of the bus routes and still cover a significant portion of its demand. The ultimate goal is to cut operational costs considerably while maintaining most of its revenue.

The park-and-ride project has two main subprojects that focus on which townships will be included as hubs, which hubs will serve which townships, and how the shuttles will operate. The objective of the latter subproject, called "shuttle subproject," is to determine the bus stops that will be served by the same shuttle, and to determine the number of shuttle buses needed to provide an acceptable service to the customers. This subproject, however, is out of the scope of this case study. In this case study, you will act as a member of the team that conducts the former subproject named as the "hub subproject." To serve as the data for the hub subproject, the Port Authority officials have provided you with the distance between each township, the demand of each township, and the capacity of possible parking sites that might serve as parking facilities. Your job is to determine which townships will include parking facilities (hubs) and which townships will be served by which hub. Note that we are making an implicit assumption that each township will be served by exactly one hub.

The current infrastructure on the buses that collect information has certain restrictions. In particular, for any given passenger, the origin-destination information is not stored in the system but rather the number of passengers getting on (similarly getting off) the bus is recorded. Due to this restriction, it is not possible to obtain and/or estimate the demand data for passengers traveling from a particular origin to a given destination. In the classical hub location problems, along with the distance matrix, demand information for the origin-destination pairs is specified. Therefore, it is not possible to utilize a classical hub location problem that makes use of travel demand between pairs of townships. Instead, it is possible to estimate the total demand for transportation at each township. Therefore, you need to use a facility location problem to model the hub subproject for the Port Authority.

Pre-analysis

Before starting the model section, answer the following questions:

1. Location problems are frequently used in practice. Find two real-life examples from newspapers, blogs, or journals within the last five years, in which benefits and/or failure of location (factory, warehouse, transportation hub) decisions made by specific companies or government agencies are discussed. What are critical conclusions in these stories?

2. Explain the advantages of using hubs instead of serving between all possible origin-destination pairs.

3. In the current project structure, the Port Authority uses two subprojects to handle hub location and shuttle assignment problems. Thus, two independent optimization problems are solved to determine these decisions. What challenges have led to this solution approach? Would it be possible to find an alternative solution approach using the same available data? What would the challenges be of solving one optimization problem that determines everything?

4. To model the hub project, we will use a modified version of one of the facility location problems that you have likely seen before during your course of study. Which facility location problem would best fit the current problem given the data at hand? Explain why and how.

   Hint: Think of a hub as a facility that will be located and townships as nodes.

Model

In the Port Authority hub subproject, we will consider setting hubs in some of the townships. We will model the problem as a modified version of the fixed-charge problem.

In this fixed-charge problem, townships will serve as nodes. In the data collected for the project, we are given a set of townships I, and the demand of each township $i \in I$ is denoted by h_i. In the context of the hub subproject, a facility is located at node $j \in J$ if a hub is located in township j. Note that the set J of townships in which a hub can be located may be different than the set of townships I. For instance, some of the townships may not be suitable for locating a hub because there is no available space to serve as a parking facility. When a facility is located at node $j \in J$, the distance of another node i ∈ I to this facility is estimated by the distance of node i to node j, which is denoted by d_{ij}. This distance gives approximate information about the distance that a customer in node i will travel to be served by a hub located at node j. Note that this is a relatively crude distance measure; we could have instead tried to simultaneously optimize the route distance, but this would add significant complexity to the problem. In addition to this added complexity, vehicle routing problems are generally operational decisions, and we only focus on the strategic decisions in this case study. We will assume that the cost per unit demand per unit distance is α, that each hub j ∈ J has a capacity C (mainly determined by the availability of parking spaces) and a fixed cost of f_j associated with locating and operating a hub to account for various administrative and rental costs.

The team for the hub subproject should determine (1) whether or not to place a facility at node j ∈ J represented by the binary variable x_j and (2) whether a node i ∈ I will be served at facility j, which is represented by the binary variable y_{ij}. The objective is to minimize total facility and transportation costs. Thus, the optimization problem for the Port Authority can be stated as follows:

$$\max \sum_{j \in J} f_j x_j + \alpha \sum_{i \in I} \sum_{j \in J} h_i d_{ij} y_{ij}$$

$$\text{s.t.} \sum_{j \in J} y_{ij} = 1 \text{ for all } i \in I \quad (1)$$

$$y_{ij} \leq x_j \text{ for all } i \in I, j \in J \quad (2)$$

$$\sum_{i \in I} h_i y_{ij} \leq C x_j \text{ for all } j \in J \quad (3)$$

$$x_j \in \{0, 1\} \text{ for all } j \in J$$

$$y_{ij} \in \{0, 1\} \text{ for all } i \in I, j \in J$$

Constraint 1 ensures that each township is served by exactly one hub. Constraint 2 states that a hub should be located at a site in order to be able to serve a township. Constraint 3 is the capacity constraint for each facility.

Data for Case Analysis

In this case study, you are asked to focus on 15 distinct townships of Allegheny County, specifically Baldwin, Hampton, Indiana, McCandless, O'Hara, Pine, North Fayette, Robinson, Ross, Shaler, South Park, Springdale, Upper St. Clair, South Fayette, and Marshall. Initial data analysis reveals that these townships have medium propensity to use public transportation, which is sufficient to cover the fixed investments of the park-and-ride project, but insufficient to keep two-way public transportation routes. Furthermore, most of the residents of these townships have cars, and they have long commuting periods, which makes these townships ideal spots for park-and-ride facilities.

You will consider each distinct township as a node. Due to the extreme cost of land and/or unavailability of the required parking space, Hampton, Pine, Robinson, and South Fayette are unsuitable for locating a parking facility. Therefore, there are only 11 locations suitable for a parking facility. The pairwise distances (in terms of miles) between each node i and each township available for hub location j are shown in Exhibit 15.1. As the table displays, the distances between two locations are not symmetric because the given distances represent actual road distances compiled as the shortest paths between the corresponding origin-destination pairs.

Clearly, the demand behavior is quite dependent on the segment of the population, as well as the time of day considered; that is, there are significant differences between regular hours versus rush (peak) hours. For the purposes of the park-and-ride project, we will focus on the rush hour behavior of the working class. Most of the transportation occurs during this time as the target population in the park-and-ride project, the working class, gets to work and returns back home. While the contribution of occasional riders during day time is not negligible, the corresponding demand is very ad hoc in nature, and thus hard to model. On the other hand, we expect to see the same type of demand pattern (albeit in the opposite direction) during both morning and evening rush hours for the working class; therefore, we will only focus on the evening rush hour (4:00 p.m.–6:00 p.m. during weekdays) for this segment of the population. The estimated average daily demand level during the evening rush hour at each node is shown in Exhibit 15.2.

In the test phase of the park-and-ride project, the Port Authority has decided to acquire and maintain hubs of uniform size; that is, each hub can serve 200 people per day. In the later phases of the project, adjusting the capacity of each hub based on the demand level will be considered. Preliminary analysis also indicated that the total daily operational cost of a hub varies for each township depending on the rental rates of parking spaces, and the corresponding estimates are shown in Exhibit 15.3. Note that this cost includes the daily cost of maintaining hubs, the cost of operating express buses, and daily wages of the drivers. It is also roughly estimated that a hub has a transportation cost of $0.10 per passenger per mile.

Study Questions

Your first and most important task is to develop and solve an appropriate facility location problem, described earlier in this chapter. You should explicitly describe this main model and give the corresponding Excel spreadsheets.

In addition to this, consider the following scenarios/conditions and explain how the model should be modified to capture these scenarios/conditions (independent of each other for each of these questions unless otherwise stated) and state the resulting optimal solution and optimal objective value in each of these cases.

1. The solution of the original model revealed that some passengers need to travel long distances to park their cars and use buses. To avoid such problems, the project team decided that the distance between a hub and a node (township) that is served by the hub cannot exceed 15 miles.
2. A hub cannot serve more than three nodes.
3. The townships of Hampton and Indiana should be served by the same hub.
4. The townships of South Park and Hampton cannot be served by the same hub.
5. Due to budgetary restrictions, in the test phase of this project, how would the solution change if you cannot locate more than three hubs? What is the opportunity cost of such a limitation; that is, how much more would you be willing to pay to be able to locate a fourth hub? What about for adding two more hubs? Determine explicitly your willingness to pay for the fourth, fifth, ..., eleventh hub. Do you think that your willingness to pay for the additional number of hubs you can have is linearly dependent on the number of hubs you are already allowed to have?
6. What happens if the transportation cost per mile per passenger changes to $0.50? How does the solution change? Specifically, how many facilities are located before and after the change in transportation cost? Explain why such a change occurs.

7. Which facility location model would you use if there were no demand data present but only the condition given in question 1? Write and solve the corresponding updated integer programming model.

8. Consider an alternative scheme. It has been decided that exactly seven hubs will be opened. For purposes of fairness, every hub will receive the same daily budget. This budget is used both for operational costs of the hub and for transportation costs of townships served by the hub. Any part of the budget that is left over is considered to be lost. What is the smallest budget possible? Write and solve the corresponding updated integer programming model.

9. What modifications would you suggest to make this model more realistic under the data restrictions explained before? What would be the corresponding parameters needed and how can those be calculated? Is there a downside of making the model more realistic?

Endnote

1. All data, characters, and project names introduced throughout the case are fictional.

Exhibits

Exhibit 15.1 Distance Matrix between Allegheny County Townships

	Baldwin	Indiana	McCand.	O'Hara	N. Fayette	Ross	Shaler	S. Park	Springdale	Up. St. Clair	Marshall
Baldwin	0	21.4	20	17.4	17.2	16.5	16.2	3.4	23.4	7.8	25.7
Hampton	21.4	5.9	5.3	8.4	25.3	6.3	5.6	24.4	12.8	23.7	12.1
Indiana	21.4	0	10.2	6	28.1	10.6	7.8	24.4	6.8	23.6	16.5
McCand.	16.7	10.2	0	11.8	22.7	4.9	7.4	23	15.8	21.6	7.4
O'Hara	14.4	6	11.8	0	23.4	8.7	4.8	20.4	10.4	19.5	18.7
Pine	21.2	12.7	5.5	15.9	26.5	9.4	11.8	27.5	17.9	26.1	4.6
N. Fayette	13.7	28	22.7	23.4	0	18.6	21.9	18.1	31.9	12.3	24.3
Robinson	10.4	21.6	14.9	16.3	8.1	10.7	14.1	16.3	24.6	13.3	19.2
Ross	12.4	10.6	4.9	8.7	18.5	0	4.5	18.8	17	17.4	10.6
Shaler	13.6	7.8	7.4	4.8	21.9	4.5	0	19.5	13.3	18.7	14.3
S. Park	7.9	24.2	23	20.4	18.1	18.9	19.2	0	26.5	7.1	28.7
Springdale	22.1	6.8	15.8	10.4	31.7	17	13.3	27.1	0	27.2	21.7
Up. St. Clair	6.4	23.6	21.6	19.9	12.3	17.5	18.7	7.3	27.2	0	27.3
S. Fayette	11	26.6	23.7	22	6.1	19.6	21.2	13.3	30.5	6.8	28.5
Marshall	22.4	16.5	7.4	18.8	24.3	10.6	14.4	28.7	21.7	27.4	0

Exhibit 15.2 Daily Demand for Each Township

Township	Baldwin	Hampton	Indiana	McCandless	O'Hara	Pine	N. Fayette	Robinson
Demand	10	92	37	142	42	58	69	67
Township	Ross	Shaler	South Park	Springdale	Upper St. Clair		S. Fayette	Marshall
Demand	156	144	67	8	96		72	35

Exhibit 15.3 Daily Operational Cost of a Hub for Each Township

Township	Baldwin	Indiana	McCandless	O'Hara		N. Fayette	Ross
Cost	200	170	190	220		230	270
Township	Shaler	South Park	Springdale	Upper St. Clair	Marshall		
Cost	190	195	180	205	260		

Part V
Analytics in Management and Ethical Decision Making

16 The Bloodgate Affair: A Case of Breaking Rules and
    Breaching Trust? 167

17 Trouble on the Thames: Event Disruption, Public Protest,
    or Public Disorder 191

16

The Bloodgate Affair: A Case of Breaking Rules and Breaching Trust?

John Davies, Victoria Business School, Victoria University of Wellington

Introduction

This case is focused on an incident that occurred in a game of rugby in 2009. Before we discuss the details of this particular incident, it is important to provide some background information about the game of rugby and the Heineken Cup competition.

Rugby

In the mid-1800s, various formats of the game of football were played under different rules in the public schools of England. Early attempts to formulate a uniform set of rules foundered as a divide established between the "dribblers" and "handlers." The former favored the Cambridge Rules and formed the Football Association in 1863. The latter favored the Rugby School rules and led to the establishment of the Rugby Football Union (of England) in 1871, the year in which the first international match in any sport, between England and Scotland, was played. By 1873, rugby union was played in Ireland, Scotland, and Wales, and together, they formed the International

Rugby Board (IRB) (now World Rugby) to regulate the increasingly popular international game.

Over the next century, rugby grew to be a global sport. It was offered in 15- and 7-a-side versions, played over 2×40 or 2×7-minute periods, respectively. Rugby was first featured in the 1924 Olympics. A quadrennial 15-a-side Rugby World Cup (RWC) was introduced in 1987. Rugby became a professional sport in 1995; and Sevens/7s is an Olympic sport for men and women as of 2016. The growth in rugby's worldwide appeal is evidenced through the 120 national teams that contest the 15-a-side Rugby World Cup (RWC), through a series of regional and continental qualifying rounds leading to a six-week finals tournament once every four years. The RWC is not only sandwiched between the years of the FIFA World Cup and the Olympic Games, but is also recognized as the third biggest global sports event behind the Olympic Games and the FIFA World Cup.

Although the global appeal of the 7-a-side game has resulted in 7s becoming an Olympic sport for the first time in Rio in 2016, the 15-a-side game, following initial lobbying from within the United States, had been featured in the 1900, 1908, 1920, and 1924 Olympics—the USA is still the existing Gold Medal holder from 1924. National teams also compete in the HSBC *World Rugby* Sevens World Series conducted in nine tournament locations in Dubai, Hong Kong, Tokyo, Las Vegas, London, Port Elizabeth, Glasgow, Australia's Gold Coast, and Wellington, New Zealand. Teams accumulate ranking points to be crowned World Series champions, and in 2015, the highest-ranking national teams qualified for the Olympic Games. There is also a quadrennial RWC for women in the 15s and 7s games and an annual World Rugby Under 20 Championship for men—a truly global game for men, women, and youth.

The Heineken Cup

The Heineken Cup, and its successor, the European Rugby Champions Cup, has been recognized, since inauguration in 1995 at the time that rugby became professionalized, as the premier competition for elite European professional rugby union teams. Twenty-four club and provincial teams from France (7 teams), England (6), Wales (4), Ireland (3), Italy (2), and Scotland (2) qualified for the Heineken Cup in 2008/2009 through high rankings in their respective English *Premiership*, the French *Top 14*, or the *Pro12* competition for Irish, Italian, Scottish, and Welsh teams. The teams are seeded in four tiers of six on the basis of past performance in European competition, so that teams in any one tier, and teams from the same country, can be separated and drawn to play in six different pools of four teams. They then play round-robin matches, with the top teams in each pool plus the two best runners-up qualifying for the knockout round of eight, the quarterfinal stage. This is the setting for the case.

Harlequins Versus Leinster

Saturday April 12, 2009—it was the final minutes of the end-of-season Heineken Cup quarterfinal clash at the Twickenham Stoop ground in London between Harlequins and the Irish provincial side, Leinster. Leinster was leading 6 points to 5—and a win would guarantee a significant payout and the promise of riches if either team progressed to the final.

The Harlequins had tried everything to score what would be winning points, but they had suffered a major setback when they lost their key man, former All Black, Nick Evans, with a knee injury just after halftime. In an earlier game against Stade Francais, he had a kicked a last-minute drop goal to give his team the victory that had taken them to the quarterfinal.

It seemed Evans's contribution could not be repeated now that he had left the field of play.

The Incident

In the 75th minute, with just 5 minutes to go, the team faced another injury and wing Tom Williams, himself a replacement in the 70th minute, needed to be replaced—with blood oozing from his mouth—by another player from the bench. That man happened to be Nick Evans!

Having left the field previously, then, in order to perform such match saving deeds again, he could only return as a blood-injury replacement.

And with 40 seconds remaining on the clock, the opportunity arose for him to pull off another match-winning trick—and he tried a 40-meter drop goal. However, the ball drifted wide, and Leinster progressed to the semifinal as one-point winners on the day. See Exhibit 16.1 for a summary of the major events in the match, especially in the second half.

At the time of the replacement, Leinster officials objected immediately—and on two counts—to Evans's return to the field. One was that he should not have been allowed to return to the field at all, having already been replaced earlier; and two, that they could see no evidence of blood on the player leaving the field.

Under the European Rugby Cup (ERC) Heineken Cup regulations, when a player is replaced, a team/club has to state whether it is a tactical or an enforced move. If the "fourth" match official is told that it's a tactical replacement, it provides scope for the player to come back on.[1]

But as Harlequins' Director of Rugby, former England captain, Dean Richards stated in response to Leinster's objections: "You have to know the rules. If they don't, it's not my problem." He also passed off Evans's return to the field as "almost a last throw of the dice," implying that while reluctant to do so, he "went with it."

Later, speaking about the loss, he said that although there were a lot of sad individuals in the changing room, that tomorrow was another day. He added casually, "There are to be no regrets," and that while the journey to the quarterfinal/round of eight was enjoyable, they should now draw a line under it—the result and the episode.[2]

However, anticipating some repercussions, former England rugby captain, BBC Radio Five-Live Commentator, and Daily Telegraph rugby correspondent Brian Moore said during the live commentary, "We will hear more about this."[3]

Indeed, the Leinster doctor, Professor Arthur Tanner, doubted the veracity of the injury, but was refused access to examine Williams in the medical room. At the post-match press conference, Leinster coach Michael Cheika also cast doubt on the nature of the injury and the appropriateness of the replacement.[4]

The Immediate Aftermath

By Friday of the following week, matters had taken another turn. Subsequent to further comments from Leinster officials and others, and widespread media coverage appearing to show Williams making a cocky wink at the bench as he left the field,[5] the Heineken Cup competition organizers, European Rugby Cup Limited (ERC), had themselves expressed concern about the second-half blood substitution of Tom Williams and fly-half Nick Evans's return to the field.

On Thursday, April 17, 2009, ERC indicated that an investigation was to be carried out by ERC disciplinary officer, Roger O'Connor, using the Heineken Cup 2008/09 disciplinary rules. His work would include gathering information from "match officials, the two teams involved and a review of broadcast footage from the game." O'Connor's brief was to determine whether any further action was necessary.[6]

A month later in mid-May, sufficient time had passed for rumors and allegations to have surfaced in the media and elsewhere that a blood capsule, previously hidden in a sock, was used to allow the switch of players to take place. O'Connor, the ERC disciplinary officer, believed there was a case to answer, and said that it would come before the ERC's independent disciplinary panel on July 1–3, 2009.[7]

The Outcome of the ERC Disciplinary Panel Hearing

Nearly two months later, Monday, July 20, 2009, following a three-day disciplinary hearing, Harlequins winger Tom Williams was found guilty of misconduct by the ERC disciplinary panel, for his role in fabricating a blood injury in the Heineken Cup quarterfinal against Leinster and was suspended for 12 months.[8]

ERC considered the matter "to be a very serious offence and one that damaged the reputation of the tournament and of rugby union."

Although the Harlequins club was also fined £215,000, with 50% suspended for two years, by contrast, the club's Director of Rugby Dean Richards and two members of the club's medical staff, physiotherapist Steph Brennan and Dr. Wendy Chapman, had misconduct complaints dismissed.[9]

Both Williams and Harlequins had the right to appeal, and Williams chose to do so. However, somewhat surprisingly, the ERC as the governing body for the Heineken Cup also decided to appeal the decision of its own independent disciplinary panel on account of the perceived unjustified leniency attributed to the Harlequins' punishment.[10]

Responses to the Verdict

Comments made by those in the rugby community, following the ERC disciplinary committee's decision, expressed surprise not only that Tom Williams, alone, had been found guilty of misconduct and at the severity of the punishment, but also surprise that the decision reflected a view that Tom Williams could have acted alone.

For example, London Irish Director of Rugby Toby Booth was surprised that Williams was singled out, but inferred the punishment showed that the authorities were perhaps attempting to send a warning to others. Former England rugby international, Mick Cleary, now a sports journalist, offered similar views, suggesting that although, on the one hand, Williams may have "taken a hit for the team," on the other hand, the ERC may have been trying to "flush out guilty pleas"—because Williams could not have made up his own mind to fake a blood injury. Cleary[11] dramatically stated that "there was not a person in the ground who did not feel that Quins were pulling a fast one when Nick Evans hobbled back into the action with five minutes remaining. People are not daft."

Expressing a player's perspective, Damian Hopley, chairman of the Rugby Players Association (RPA), viewed both the decision and punishment as extraordinary given Williams's exemplary record and in light of recent relatively lesser punishments meted to others involved in physical foul play and drug taking.[12]

The view of the Harlequins club was conveyed as surprise and disappointment at the Williams guilty verdict, especially so because others—Dean Richards, Steph Brennan, and Dr. Wendy Chapman—were acquitted on similar charges.[13]

The Delayed Aftermath

In the months that followed the initial verdict, the Harlequins club conducted an internal enquiry into the incident. The ERC Disciplinary Panel also heard and ruled on appeals of its initial decision in the matter.

Harlequins Internal Enquiry

Saturday, August 8, 2009—Just two weeks later, following an internal enquiry within the Harlequins club, and following the ERC ruling, Director of Rugby Dean Richards resigned.

Harlequins CEO, Mark Evans, in a statement on the club website said:

> We have been found guilty of behaviour that cannot be accepted or condoned. For that we apologise to you unconditionally.

He added a view that not only would the decision drive the club to ensure that the highest standards would be upheld in the future, but that as a consequence, every other professional rugby club would be reviewing their own practices as they learn from what happened.

Appeals to the ERC Disciplinary Panel—Monday, August 17, 2009

Following a 14-hour disciplinary appeal hearing, Dean Richards was banned for three years from involvement in ERC-affiliated rugby. At the appeal, Tom Williams changed his previous evidence, and provided details of how Richards and physiotherapist Steph Brennan, by then employed by the England team, "colluded to fake a blood injury in the game and then orchestrated a cover-up."[14]

Williams had alleged that he was cut in the mouth by a club official as part of an attempt to make the fake injury appear real—someone later identified as team doctor and A&E specialist, Wendy Chapman.[15]

Brennan was banned for two years from participation in all European rugby sanctioned by the ERC after pleading guilty and admitting his role in the cover-up. He was later sacked by the England RFU.[16]

By turning Queen's Evidence and exposing the nature of the Harlequins deception, Williams had his original ban reduced to four months. Harlequins had their original fine of €250,000, half of which was suspended, increased to €300,000 (£258,000) with none suspended, but were not banned from the Heineken competition as had been expected.

While the ERC could only apply bans within Europe, they urged other national governing bodies worldwide to adopt the suspensions.

ERC Appeal Committee Chairman Rod McKenzie said details of four other occasions on which Harlequins had attempted to fake injuries had been passed to the relevant authorities.[17,18]

Response to the Appeal Decisions

Richards said he accepted responsibility for the blood-capsule incident and for ordering the "blood substitution" but professed ignorance of the cut administered to Williams until some eight days later. He was also disappointed and surprised at the length of his ban.

He indicated that following the match, his later actions had been motivated by a sense of managerial duty and sense of loyalty "to safeguard the professional position of those involved" and the "mistaken belief" that he was acting in the interests of the club. Consequently, for example, and with the agreement of Tom Williams, he had suggested the fabrication of a story, withholding the true facts about the substitution. He also admitted that doing so was "obviously wrong," and apologized for his actions, and the subsequent damaging publicity that impacted Harlequins, its staff, his own family, and the game.[19]

Damian Hopley, the RPA CEO, expressed gratitude that Williams's 12-month ban had been reduced on appeal, yet accepted that the scandal had left "an indelible stigma on the professional game." He restated his view that while Williams was "a young man of good character," he had made "a serious error of judgement." His view was that the players have an increasing responsibility to act as role models for the sport and should take a leading role in restoring the sport's "damaged image and integrity."[20] He regarded the decisions as a strong deterrent to similar unethical behavior and expressed the players' view there was no room in the game for cheating. He claimed, however, that for a deterrent to be effective, all disciplinary sanctions handed down in rugby, including those arising from cross-border competitions, must be applied by all unions in membership of the IRB—in keeping with the IRB Regulations and "the application of the universality principle." That is, a disciplinary decision made by one governing body should be respected by all—for example, to ensure that a player banned in one country, competition, or jurisdiction should be banned in all.

On August 24, 2009, the Rugby Football Union disciplinary officer, Judge Jeff Blackett, stated that he had been shaken by the incident and subsequent events. In his view, the subsequent events involving a doctor allegedly cutting a player in the mouth to camouflage the original deception and the willingness of a number of respected people to lie to a disciplinary panel were more serious than the cheating on the pitch.

He had always believed that rugby people have been "pretty honest," prepared to own up to their misdeeds and take their medicine. He had been shocked that Dean Richards, who had been regarded as "the essence of English rugby" when he played and as a coach, and who was someone he had trusted when he represented his players at disciplinary hearings, had lied. He pondered that if a man like Richards can succumb to pressures in the professional game, "where does that leave rugby football?"

Within a week, August 28, 2009, the Harlequins chairman, Charles Jillings, who had been implicated by Williams as having tried to craft his testimony at the Appeal Hearing, resigned.[21]

Epilogue

One year later, August 2010, Tom Williams was back playing for Harlequins. Richards had been given clearance to work on a one-off basis as a rugby "consultant" with Worcester rugby. At the same time, Brennan and Chapman continued to be under temporary suspension awaiting trial by their respective professional councils, the Health Professions Council (HPC) and the General Medical Council (GMC).

In September 2009, Dr. Wendy Chapman had been suspended from practicing by the General Medical Council (GMC) and left her job as an Accident & Emergency consultant, having admitted to deliberately cutting the lip of Tom Williams as part of an attempt to conceal that Williams had bitten into a blood capsule.[22] Chapman had initially denied her role in the incident to the ERC Disciplinary Panel, was acquitted, but later recanted and gave evidence for the prosecution at a later appeal hearing, expressing shame and embarrassment about her "appalling lack of judgement" in the face of "huge pressure."[23]

One year later, in September 2010, she faced a GMC Disciplinary Panel with the possibility of being permanently struck off the medical register. However, the GMC found that although she did not act in Williams's best interest at the time, and had acted under pressure from Williams, that her ability to practice was no longer impaired and ruled that she be reinstated and be free to work in medicine again.[24]

Following the ERC Disciplinary Hearing, Steph Brennan, the club physiotherapist and recent appointment to the England team, was banned by the ERC from all rugby competitions under their jurisdiction and was suspended from his role with England pending formal employment review processes, but later resigned. One year later, he appeared before a Health Professions Council misconduct hearing and was struck off—that is, banned from practicing—for life.[25]

It was ruled that Brennan had known of, organized, and/or assisted in the fabrication of a blood injury to Tom Williams during the match against Leinster, and that he had purchased fake blood capsules and provided a fake capsule to Williams in an attempt to cheat. It was also ruled that he had attempted to conceal the fake blood injury to Williams, had provided untruthful and/or inaccurate evidence during the course of an ERC disciplinary hearing, and had been involved in fabricating blood injuries in games on a number of occasions.

Brennan had admitted his role in the affair and had expressed remorse, stating, "I followed orders and wish I hadn't. Yes, I went on to pitch with the intention of deceiving the referee." It was reported that a fear of having his contract terminated left him with little option but to obey Richards's previous requests to use blood capsules. However, the ruling stated his behavior had been "dishonest, premeditated and ... continued over a considerable period of time starting in the 2005/2006 season, and merited a life ban."[26]

The ERC appeals process meant that Richards, Brennan, and Williams had no further right to appeal their punishments with the ERC, but that any further appeal could be taken to the Court of Arbitration for Sport—which they chose not to do.

The Task

You've been brought in to advise the new Harlequins CEO about the club's responsibilities to promote ethical behavior among those producing and delivering its major product—a game of rugby—and then to develop an organizational ethos supportive of such behavior.

You have considered, like others, introducing an ethics education program. But you've been advised to ensure that in making a case for such a program that you need to build a convincing argument using the findings of an in-depth stakeholder analysis. In short, you need to persuade the doubters that a spectrum of stakeholders are likely to be affected by, and react in different ways to, ethical breaches—and that this means you need to take account of emotional responses as well as other rational economic responses.

You've also taken on board advice to ensure that the players should understand how moral values and moral emotions can shape behavior and that players should also develop an ability to distinguish between different levels of moral reasoning exhibited by stakeholders, including players. It has also been suggested that players, in particular, should be sensitized to how unthinking choices and actions can lead to a slippery slope of escalating commitment to continue those actions, so that understanding the consequences of early choice is important.

Finally, in getting a program off the ground, you know that the club has to understand how its programs contribute to wider initiatives being developed by the governing bodies, the Rugby Union of England, and the European Rugby Cup group and that once it puts an ethical code in place, it needs to understand how its authority to act and its area of jurisdiction complement those of other governance groups.

The analysis that we agreed upon is set out more fully in the following study questions. Best of luck!

Study Questions

1. Why has the case attracted such media attention?

   a. Describe why this case should have created so much public interest and attracted such attention in the sports world.

   b. Comment on the social context, the athletes, the sport, the fans, and the media.

   c. Outline the events associated with the affair in chronological order; describe Williams's escalation of involvement, and identify moments when he could have made different decisions.

   d. Draw parallels with examples of unethical behavior that have targeted the operations of organizations and sports events.

2. Stakeholders and Stakeholder Analysis: Outline the major "players" involved in, or affected by, the Bloodgate Affair—that is, identify the key stakeholders relating to the Bloodgate Affair.

   a. Describe their main attributes, responsibilities, stakes, and interests.

   b. Describe how the stakeholders impacted and were impacted by public perception of the sports event and the public protest.

   c. Provide a chart of generic stakeholders of the Harlequins club and state how they may have been impacted by the actions of the key players, such as Tom Williams (player), Dean Richards (director of rugby), Mark Evans (Harlequins CEO), Steph Brennan (team physio), and Wendy Chapman (team doctor). Indicate two specific stakeholders who can be labeled as dangerous or dominant stakeholders.

3. Ethical Dilemmas and Ethical Behavior:

Chronology of Ethical Dilemmas

Williams, Richards, and later Mark Evans would each have faced a series of ethical dilemmas as the case unfolded.

  a. Outline the particular chronology of ethical dilemmas faced by Williams, Richards, and Evans and the decisions that they took.

  b. Consider what factors may have given rise to the unethical behavior. (Use Cressey's Fraud Triangle to assess motivation/pressure, opportunity, and rationalization for such behavior.)

Moral Values and Moral Emotions

  c. What moral values were breached by Tom Williams and Dean Richards in their contributions to the Bloodgate Affair? (Use Lumpkin, Beller, and Stolle's Conceptualization of Moral Values.)

  d. What moral values may have been breached by the ERC Disciplinary Panel in their first decisions to fine the Harlequins club, but only one individual, Tom Williams?

  e. Provide some examples of the moral emotions manifested in the media coverage of the affair. (Use Haidt's notions.)

  f. Describe the moral emotions that appear to have surfaced in those commenting on events, and how those emotions may impact moral judgments and decisions taken by the different actors.

Moral Reasoning

  g. Using Kohlberg's framework, indicate the level of moral reasoning exhibited by Tom Williams and Mark Evans for each of the dilemmas that you have identified.

  h. Briefly indicate how Tom Williams and Mark Evans (the player and CEO) could have managed those situations differently if they had recognized the successive dilemmas.

4. Governance, Multiple Jurisdictions, Double Jeopardy, Justice and Fairness, and Locus of Responsibility
   a. Offer some brief comment on where the locus of responsibility should lie in the Bloodgate Affair. Should it lie with the players, coach, medical team, club management, England RFU, the ERC as Heineken Cup organizers, the IRB, and/or the GMC and HPC as professional bodies?

   That is, offer reasoned comment on who should take responsibility for promoting ethical behavior.

5. Applicability to other Managerial or Professional Situations
   a. What are the implications, the ethics lessons, from the Bloodgate Affair for those involved in management and governance?
   b. Can the situation be seen as presenting opportunities for the key people to accept responsibility for leadership, values, and ethics; to be honest; and to convey integrity?
   c. Consider whether and why cheating or fraudulent behavior is more or less acceptable in some contexts than in others.
   d. If cheating is less acceptable in a professional environment, what actions could sport managers and/or sport governing bodies take to reduce the likelihood of cheating?

Endnotes

1. Interestingly, during the subsequent Heineken Cup Semifinal against Munster, Leinster's Felipe Contepomi left the field on a stretcher, but the fourth official was told it was a tactical replacement, leaving scope for him to return to the field of play.

2. M. Cleary. 2009a. "Leinster Scrape Past Harlequins into Heineken Cup Semi-Finals," *The Daily Telegraph*, April 13, 2009, http://www.telegraph.co.uk/sport/rugbyunion/club/5145433/Leinster-scrape-past-Harlequins-into-Heineken-Cup-semi-finals.html.

M. Cleary. 2009b. "Tom Williams Ban Branded 'Excessive and Entirely Disproportionate' by Players Union," *The Daily Telegraph*, July 21, 2009, http://www.telegraph.co.uk/sport/rugbyunion/club/5880985/Tom_Williams-ban-branded-excessive-and-entirely-disproportionate-by-players-union.html.

ERC, P. Griffin. 2009a. "Dean's Dream in Ruins after Leinster Loss," *ERC Media Release*, April 12, 2009, http://www.ercrugby.com/eng/2092.php.

3. B. Moore. 2009b. "Harlequins's 'Bloodgate' Covers Nobody in Glory," *The Daily Telegraph*, August 10, 2009, http://www.telegraph.co.uk/sport/rugbyunion/club/6000284/Harlequinss-bloodgate-covers-nobody-in-glory.html.

4. P. Kelso. 2009b. "Dean Richards Given Three-Year Coaching Ban after Harlequins Bloodgate Scandal," *The Daily Telegraph*, August 18, 2009, http://www.telegraph.co.uk/sport/rugbyunion/club/6045603/Dean-Richards-given-three-year-coaching-ban-after-Harlequins-bloodgate-scandal.html.

5. B. Moore. 2009b. "Harlequins's 'Bloodgate' Covers Nobody in Glory," *The Daily Telegraph*, August 10, 2009, http://www.telegraph.co.uk/sport/rugbyunion/club/6000284/Harlequinss-bloodgate-covers-nobody-in-glory.html.

6. BBC, "Quins' Substitution Under Scrutiny," *BBC*, April 17, 2009, from http://news.bbc.co.uk/sport1/hi/rugby_union/my_club/harlequins/8005131.stm.

7. P. Ackford, "Rugby in Tune with the Week of Shame," *The Daily Telegraph*, May 16, 2009, http://www.telegraph.co.uk/sport/rugbyunion/club/5331385/Rugby-in-tune-with-the-week-of-shame.html.

8. ERC, J. Corcoran. 2009b. "Harlequins Misconduct Complaints Hearing," *ERC Media Release*, July 1, 2009, http://www.ercrugby.com/eng/6465.php.

    ERC, A. Lacroix. 2009c. "Harlequins Misconduct Complaints Hearing Adjoined," *ERC Media Release*, July 3, 2009, http://www.ercrugby.com/eng/6462.php.

    ERC, J. Corcoran. 2009d. "Harlequins Misconduct Complaints Hearing Decision," *ERC Media Release*, July 20, 2009, http://www.ercrugby.com/eng/6473.php.

9. Telegraph. 2009a. "Harlequins' Tom Williams Banned for 12 Months for 'Faking an injury,'" *The Daily Telegraph*, July 20, 2009, http://www.telegraph.co.uk/sport/rugbyunion/club/5873901/Harlequins-Tom-Williams-banned-for-12-months-for-faking-an-injury.html.

10. ERC, P. Griffin. 2009f. "Misconduct Hearing Appeals," *ERC Media Release*, August 8, 2009, http://www.ercrugby.com/eng/6556.php.

11. M. Cleary. 2009b. "Tom Williams Ban Branded 'Excessive and Entirely Disproportionate' by Players Union," *The Daily Telegraph*, July 21, 2009, http://www.telegraph.co.uk/sport/rugbyunion/club/5880985/Tom_Williams-ban-branded-excessive-and-entirely-disproportionate-by-players-union.html.

12. M. Cleary. 2009b. "Tom Williams Ban Branded 'Excessive and Entirely Disproportionate' by Players Union," *The Daily Telegraph*, July 21, 2009, http://www.telegraph.co.uk/sport/rugbyunion/club/5880985/Tom_Williams-ban-branded-excessive-and-entirely-disproportionate-by-players-union.html.

13. Telegraph. 2009a. "Harlequins' Tom Williams Banned for 12 Months for 'Faking an injury,'" *The Daily Telegraph*, July 20, 2009, http://www.telegraph.co.uk/sport/rugbyunion/club/5873901/Harlequins-Tom-Williams-banned-for-12-months-for-faking-an-injury.html.

14. ERC, P. Griffin. 2009f. "Misconduct Hearing Appeals," *ERC Media Release*, August 8, 2009, http://www.ercrugby.com/eng/6556.php.

    ERC, A. Lacroix. 2009g. "Appeals Hearing," *ERC Media Release*, August 11, 2009, http://www.ercrugby.com/eng/6549.php.

    ERC, J. Corcoran. 2009h. "Misconduct Appeal Hearings," *ERC Media Release*, August 17, 2009, http://www.ercrugby.com/eng/6538.php.

    ERC, R. O'Connor. 2009i. "Decision of Appeal Committee in Appeal," *ERC Disciplinary Officer*, August 17, 2009, http://www.ercrugby.com/AR-M700U_20090902_085314.pdf.

    ERC, Tom Williams. 2009j. "Decision of Appeal Committee in Appeal, Misconduct Appeal Hearing Decisions," *ERC Media Release*, August 17, 2009, http://www.epcrugby.com/news/6538.php.

    ERC, P. Griffin. 2009k. "Williams Regrets 'Grave Error of Judgement,'" *ERC Media Release*, August 18, 2009, http://www.ercrugby.com/eng/6542.php.

    ERC, P. Griffin. 2009. "Independent Appeal Committee Decision—Tom Williams," *ERC Media Release*, August 25, 2009, http://www.ercrugby.com/eng/6547.php.

    C. Hewett, "Doctor at Centre of 'Bloodgate' Admits Role in Cutting Player's Lip," *The Independent*, August 24, 2010, http://www.independent.co.uk/sport/rugby/rugby-union/news-comment/doctor-at-centre-of-bloodgate-admits-role-in-cutting-players-lip-2060158.html.

15. P. Kelso. 2009a. "Tom Williams: I Was Cut in Harlequins Bloodgate Cover-Up," *The Daily Telegraph*, August 14, 2009, http://www.telegraph.co.uk/sport/rugbyunion/club/6031583/Tom-Williams-I-was-cut-in-Harlequins-bloodgate-cover-up.html.

16. ERC, P. Griffin. 2009k. "Williams Regrets 'Grave Error of Judgement,'" *ERC Media Release*, August 18, 2009, http://www.ercrugby.com/eng/6542.php.

17. As long ago as 2001, a former England coach and player, and Leicester coach, Richard Cockerill, stated that fake blood injuries were occurring during Premiership games. He also admitted that stitches in his finger were ripped to fake a blood injury.

18. C. Hewett, "Doctor at Centre of 'Bloodgate' Admits Role in Cutting Player's Lip," *The Independent*, August 24, 2010, http://www.independent.co.uk/sport/rugby/rugby-union/news-comment/doctor-at-centre-of-bloodgate-admits-role-in-cutting-players-lip-2060158.html.

19. P. Kelso. 2009b. "Dean Richards Given Three-Year Coaching Ban after Harlequins Bloodgate Scandal," *The Daily Telegraph*, August 18, 2009, http://www.telegraph.co.uk/sport/rugbyunion/club/6045603/Dean-Richards-given-three-year-coaching-ban-after-Harlequins-bloodgate-scandal.html.

    P. Rees. 2009a. "'I Ordered Fake Blood Substitution,' Admits Disgraced Dean Richards," *The Guardian*, August 19, 2009, http://www.guardian.co.uk/sport/2009/aug/19/dean-richards-denial-cut-tom-williams.

20. Guardian, "Dean Richards Scandal Has Left Indelible Stain on the Game, Says PRA," *The Guardian*, August 18, 2009, http://www.guardian.co.uk/sport/2009/aug/18/dean-richards-damian-hopley.

21. ERC, J. Corcoran. 2009h. "Misconduct Appeal Hearings," *ERC Media Release*, August 17, 2009, http://www.ercrugby.com/eng/6538.php.

22. P. Kelso. 2009a. "Tom Williams: I Was Cut in Harlequins Bloodgate Cover-Up," *The Daily Telegraph*, August 14, 2009, http://www.telegraph.co.uk/sport/rugbyunion/club/6031583/Tom-Williams-I-was-cut-in-Harlequins-bloodgate-cover-up.html.

    C. Hewett, "Doctor at Centre of 'Bloodgate' Admits Role in Cutting Player's Lip," *The Independent*, August 24, 2010, http://www.independent.co.uk/sport/rugby/rugby-union/news-comment/doctor-at-centre-of-bloodgate-admits-role-in-cutting-players-lip-2060158.html.

23. ERC, J. Corcoran. 2009h. "Misconduct Appeal Hearings," *ERC Media Release*, August 17, 2009, http://www.ercrugby.com/eng/6538.php.

    ERC, Roger O'Connor. 2009i. "Decision of Appeal Committee in Appeal, ERC Disciplinary Officer," August 17, 2009, http://www.ercrugby.com/AR-M700U_20090902_085314.pdf.

    ERC, Tom Williams. 2009j. "Decision of Appeal Committee in Appeal, Misconduct Appeal Hearing Decisions," *ERC Media Release*, August 17, 2009, http://www.epcrugby.com/news/6538.php.

24. J. Lawton, "James Lawton: The Bloodgate Physio Is Banned for Life. So Why Is the Door Open for Richards' Return?" *The Independent*, September 18, 2010, http://www.independent.co.uk/sport/rugby/rugby-union/news-comment/james-lawton-the-bloodgate-physio-is-banned-for-life-so-why-is-the-door-open-for-richards-return-2082456.html.

    Telegraph, "Bloodgate Doctor Wendy Chapman 'Did Not Act in Patient's Best Interests,'" *The Daily Telegraph*, August 26, 2010, http://www.telegraph.co.uk/sport/rugbyunion/club/7965804/Bloodgate-doctor-did-not-act-in-patients-best-interests.html.

25. J. Lawton, "James Lawton: The Bloodgate Physio Is Banned for Life. So Why Is the Door Open for Richards' Return?" *The Independent*, September 18, 2010, http://www.independent.co.uk/sport/rugby/rugby-union/news-comment/james-lawton-the-bloodgate-physio-is-banned-for-life-so-why-is-the-door-open-for-richards-return-2082456.html.

    G. Mairs, "Harlequins' 'Bloodgate' Physio Steph Brennan Begins High Court Appeal to Save Career," *The Daily Telegraph*, December 14, 2010, http://www.telegraph.co.uk/sport/rugbyunion/club/8201631/Harlequins-bloodgate-physio-Steph-Brennan-begins-high-court-appeal-to-save-career.html.

26. J. Lawton, "James Lawton: The Bloodgate Physio Is Banned for Life. So Why Is the Door Open for Richards' Return?" *The Independent*, September 18, 2010, http://www.independent.co.uk/sport/rugby/rugby-union/news-comment/james-lawton-the-bloodgate-physio-is-banned-for-life-so-why-is-the-door-open-for-richards-return-2082456.html.

    G. Mairs, "Harlequins' 'Bloodgate' Physio Steph Brennan Begins High Court Appeal to Save Career," *The Daily Telegraph*, December 14, 2010, http://www.telegraph.co.uk/sport/rugbyunion/club/8201631/Harlequins-bloodgate-physio-Steph-Brennan-begins-high-court-appeal-to-save-career.html.

    B. Moore, "Steph Brennan, Former Harlequins Physio, Unfairly Struck-Off by Health Council," *The Daily Telegraph*, September 15, 2010, http://www.telegraph.co.uk/sport/rugbyunion/club/8005099/Steph-Brennan-former-Harlequins-physio-unfairly-struck-off-by-health-council.html.

Additional Reading

ERC. 2009e. "Tom Williams and Harlequins Independent Disciplinary Committee Decision," *ERC Media Release*, July 2, 3, and 20, 2009, http://www.ercrugby.com/images/content/Tom_Williams_and_Harlequins_Independent_Disciplinary_Committee_Decision.pdf.

Moore. 2009a. "Tom Williams Made the Fall Guy at Harlequins in Misguided ERC Crackdown," *The Daily Telegraph*, July 23, 2009, http://www.telegraph.co.uk/sport/rugbyunion/club/5888401/Tom-Williams-made-the-fall-guy-at-Harlequins-in-misguided-ERC-crackdown.html.

P. Rees. 2009b. "Dean Richards Turns Down Leicester Legends Parade," *The Guardian*, September 17, 2009, http://www.guardian.co.uk/sport/2009/sep/17/dean-richards-leicester-legends-bloodgate.

Telegraph. 2009b. "Harlequins Plea to Stay in Heineken Cup," *The Daily Telegraph*, August 12, 2009, http://www.telegraph.co.uk/sport/rugbyunion/club/6018930/Harlequins-plea-to-stay-in-Heineken-Cup.html.

Exhibit

Exhibit 16.1 How the Match Unfolded

April 12, 2009 Heineken Cup knock-out quarterfinal match, Harlequins versus Leinster at the Twickenham Stoop ground in SW London.

First Half

In a match characterized by physical confrontation, and described as a physical encounter, Harlequins' key player and goal kicker, Nick Evans, the former All Black, was subject to a number of big hits or tackles, including one from Leinster's Australian flanker, Rocky Elsom.

Leinster led by 6 points to nil, having kicked two penalty goals (akin to field goals in the NFL).

Second Half

47 mins—Evans's body "gives way" and he "hobbles off" with a knee injury to make way for another reliable goal kicker, Chris Malone.

66 mins—Malone provides a major assist in a try (5 points) scored by Harlequins' England full-back Mike Brown.

Leinster still leads by 6 points to 5.

67 mins—Malone attempts a 2-point conversion kick, but misses.

69 mins—Malone "hobbles off" with a hamstring injury. Harlequins has no further reliable kicker. Instead, a substitute utility player, Tom Williams, enters the field as a replacement/substitute.

71 mins—Harlequins awarded a penalty kick at goal, for an infringement by Leinster at a scrum (akin to a scrimmage). Mike Brown fails to kick the goal, which would have given Harlequins the lead with 9 minutes remaining in the match.

75 mins—Williams leaves the field with blood coming from what seems to be a blood injury.

As he leaves the field, Williams is seen winking toward the Harlequins bench as Nick Evans prepares to return to the game.

76 mins—Leinster officials make an immediate protest/query to referee Nigel Owens, about the validity of Evans's return. Owens talks to Harlequins coach Dean Richards, who confirms that Evans is returning to the field in a substitution for a blood injury and not as a tactical substitution.

79 mins—Evans, who was a late drop-goal hero against Stade Francais in an earlier pool stage match of the Heineken Cup, misses a "drop-goal" attempt to score 3 points.

80 mins—Leinster hangs on for victory.

Data source: Derived mainly from:

M. Cleary. 2009a. "Leinster Scrape Past Harlequins into Heineken Cup Semi-Finals," *The Daily Telegraph*, April 13, 2009, http://www.telegraph.co.uk/sport/rugbyunion/club/5145833/Leinster-scrape-past-Harlequins-into-Heineken-Cup-semi-finals.html.

P. Kelso. 2009b. "Tom Williams: I Was Cut in Harlequins Bloodgate Cover-Up," *The Daily Telegraph*, August 14, 2009, http://www.telegraph.co.uk/sport/rugbyunion/club/6031583/Tom-Williams-I-was-cut-in-Harlequins-bloodgate-cover-up.html.

17

Trouble on the Thames: Event Disruption, Public Protest, or Public Disorder

John Davies, Victoria Business School, Victoria University of Wellington

Introduction

The Boat Race is a highlight event on Britain's sporting calendar, with a standing and importance in the sporting psyche of the nation that denies it being just a race between two amateur crews of eight, plus a cox, from Oxford and Cambridge universities, also known as Oxbridge. The race has taken place on the River Thames almost without break since 1845 on a winding 4 miles, 374 yards (6.8km) course up river between Putney Bridge and Chiswick Bridge, Mortlake, in South West London.[1] (The race is timed to start on the incoming flood tide, an hour before high tide, so that the crews are rowing with the fastest possible current.) The Boat Race started as a private challenge between undergraduate friends on the river at Henley in 1829.[2] Nowadays, the Boat Race is organized by the Boat Race Company Ltd. (BRCL) and receives a multimillion dollar title sponsorship from the BNY Mellon investment management organization and others,[3] and starting in 2014, Nike is the official Performance Wear Partner.

Spectators, many of whom have little or no involvement in rowing, will line the 8 miles of riverbank and Putney, Hammersmith, Barnes, and Chiswick Bridges to catch a glimpse of the event, or just to say "I was there!" The event is not without drama, and weather conditions and choppy tides have sometimes resulted in boats capsizing! The event was one of the first to be covered as an outside broadcast (by the BBC) and has been a landmark media event since the first days of mass TV broadcasting in the 1950s. The Boat Race is also a major social event providing opportunity for "old boys" reunions and later for the sponsors like BNY Mellon to use the Boat Race and associated hospitality functions to build relationships with business partners.

Whereas at one time the crews were entirely made up of British students, in recent years the prestige of the event and the gaining of a "blue"—dark blue for Oxford and light blue for Cambridge—has attracted scholar athletes to Oxbridge from the Americas, Europe, Australasia, and Africa. Many of these athletes set their "blues" alongside their Olympic and World Championship rowing medals in terms of pride and achievement.

The Act of Protest at the Oxford-Cambridge Boat Race

On Saturday, April 8, 2012, with just over half of the 158th annual Boat Race completed, Trenton Oldfield jumped into the Thames to swim in front of and between the boats as they approached the Chiswick Eyot. The race was immediately stopped for safety reasons by the assistant umpire, former Olympic gold medalist rower, Sir Matthew Pinsent.

The race was restarted 30 minutes later at the eastern end of the Chiswick Eyot by Umpire John Garrett with unfortunate consequences. The boats collided, and the oar of Oxford rower Hanno Wienhausen was broken, putting Oxford at a considerable disadvantage.

However, Umpire Garrett let the race continue, allowing Cambridge to take the win by the surprisingly small margin of 4 lengths—a victory that was appealed but allowed to stand.

The disruption and halt to the race meant that it was the first time since 1849 that a crew had won the Boat Race without an official recorded winning time. Not surprisingly, the postrace celebrations of the Cambridge crew were muted, and the awards ceremony was canceled.

The Protester

At the time of his protest, Trenton Oldfield was 35 years old, living in a small flat in the East End of London with his wife, Deepa Naik. Together with his wife, he ran an independent publishing house, organized urban festivals, ran two not-for-profit organizations promoting community involvement in urban policy development, and was engaged as a community worker.[4]

He had grown up in Sydney, Australia, and had been educated at one of Australia's exclusive private schools, the Sydney Church of England Grammar School, leaving at 16 "because I couldn't stand the elitism." He eventually gained a scholarship to study at LSE—the London School of Economics—working his way through his studies and completing an MSc in contemporary urbanism. It was there he became disturbed by visible inequalities in society. "I protest their injustices—ask anyone that knows me."[5]

At the time of the Boat Race, Oldfield had just returned from six months in Canada following a period of caring for his wife's terminally ill father. In the week before his protest, he had been upset by legislation that he perceived would lead to privatizing the NHS, by the introduction of the Data Communications Act, and by a government minister urging the public to report their neighbors if "they suspected them of planning to protest at the Olympics." He implied that these events had been the final straw for him and that they had led to the idea of using the Boat Race for protest; "the next day, I went out and bought the wetsuit."[6]

This is not the first time that a protestor had used a sporting event to deliver his or her message. See Exhibit 17.1 for a summary of major protests at sporting events that have occurred since 1900.

The Immediate Aftermath

As Oldfield was being lifted out of the water by the river police lifeboat crew, he said he had no sense or feeling of threat and that people were joking and laughing. He implied that even the police were low key, with their understated question: "What was all that about then?"

Indeed, he appeared to have a broad smile on his face—before police arrested him on suspicion of a public order offence and took him into custody.[7]

He was in custody overnight unaware that "Twitter was hissing with fury and death threats,"[8] and he was surprised at the media attention received when he was released the following morning.

He claimed that his protest arose from a feeling of "heartbreak" at the deepening inequality in British society, and was a protest against "elitism and inequality," the growing culture of elitism demonstrated by the spending cuts of the coalition government, and the erosion of civil liberties.[9]

The Reaction

This section discusses the reactions to the protest offered by various constituencies.

The Media

Renowned journalist Rupert Myers[10] offered a set of balanced views—for example, that although Oldfield may have been misguided, he had clearly never sought to hurt others or to gain financially from his protest. He also implied a contradiction in that although Oldfield disapproved of elitism, he had chosen to disrupt a meritocratic contest. He considered that Oldfield had acted in the manner of the dumbest protesters, ill-thought through, aimed at annoying Oxbridge types.

The Athletes/Rowers

Martin Cross, the 1984 Olympian, readily accepted that rowing is elitist, but not as Oldfield claimed it was—that is, based on social class, but one based on elite athletes pursuing performance-based success. Past and present rowers offered more damning opinions on events. For Karl Hudspith, OUBS president in 2012, it was a matter of taking away the climax to the student athletes' rowing careers, for which they had prepared with "seven months of hell." For David Nelson, CUBC president, Oldfield's actions were seen as callous and selfish, negating the efforts of students who had tested themselves to the limit in tough training regimens for no other reward than "challenge, team-spirit and the deep satisfaction of personal achievement" and winning! Cross regarded Oldfield's actions as reckless, no more than a stunt, but still a "misguided assault" on those core values of challenge and so forth.[11]

By comparison, William Zeng, OU rower, formerly of Yale, and a Rhodes Scholar, completing a DPhil in computer science at Oriel College, was scathing in his comment that Oldfield was "a mockery of a man," who would make a mockery of the rowers' dedication and courage. He suggested that Oldfield was, in effect, protesting the rowers' right to demonstrate hard work and desire in a proud tradition and protesting their right to pursue "the joys and hardship of sport."[12]

The Officials/Organizers

David Searle, chief executive of the Boat Race Company, also referred to the protest as a stunt, saying that there was little anyone could have done to prevent an individual "from staging such a stunt." Race Umpire John Garrett also commented on averting such threats to the event, and told the BBC that, although the possibility of swimmers had been discussed the previous year, the threat had not been expected in 2012.[13]

Focusing on the swimmer, the four-time Olympic gold medalist and assistant umpire, Sir Matthew Pinsent, prepared a statement for the trial, outlining his concern that "the risk to the swimmer was great," and that Oldfield could have been killed. He suggested that if the rowers had not raised their oars to miss him, he could have "cracked his skull on the metal rigging, been knocked unconscious and drowned."[14]

Oldfield's Reaction

Oldfield had previously made comments disputing claims that he had put himself in danger. He said that having grown up in Australia, he was used to "dodging surfboards, rock and boats."[15] In response to other criticisms, Oldfield used Twitter to comment in a mocking manner that he expected nothing other than for the "vindictive" class to be vindictive about having their fun and their "months of training" disrupted, and in a defiant manner that he was prepared to go to jail for his beliefs.[16]

The Judiciary and the Verdict

In setting the scene for the trial, Prosecutor Louis Mably implicitly paralleled the arguments offered by others in claiming that the Boat Race had been spoiled for "hundreds of thousands of spectators," as well as the two crews.

The Isleworth Court agreed, and Oldfield, having been charged under section 5 of the Public Order Act, was convicted in October 2012 of causing a public nuisance. He was fined £750, sentenced to six months in jail, and later had his spousal visa application to remain in the UK rejected by the Home Office.[17]

At his trial conducted in September 2012, Judge Anne Molyneux reflected on both the positives and negatives of the man and his actions. She said, for example, following the verdict that he was a man attested to be of good character, who had contributed to the community through his employment and demonstrated several desirable qualities.[18] However, she also said that he had taken planned, deliberate, and disproportionate action that not only exhibited prejudice, but also was dangerous and had no regard for the sacrifices the rowers had made.[19]

Subsequently, the Home Office denied Oldfield's application for a spousal visa. Their stance was that "those who come to the UK must abide by our laws"—their rationale being a belief that his presence in this country would not be conducive to the public good.[20]

Oldfield's Supporters

Oldfield's supporters believed that he had been victimized by the system—perhaps on the basis that his charge had been elevated from a public order offence to public nuisance, the latter having potential for a custodial sentence with a longer term of imprisonment.[21] In general, though, "support" was couched more as criticism of the stances taken by authorities more than support for Oldfield's views.

For example, Mitch Mitchell, a member of the campaign group, *Defend the Right to Protest,* had formed the view that "the authorities" were cracking down harder on those "who raise a voice of protest." The respected, veteran journalist campaigner, John Pilger, took a stronger view and interpreted the Home Office decision as implying a stance that could potentially criminalize all forms of protest.

However, in Pilger's judgment, Oldfield was neither a criminal nor a terrorist. He credited Oldfield with being a *protester* and acting on principle, submitting that this was so, regardless of whether people agreed with his actions.

Epilogue

Following early release from Wormwood Scrubs prison, Oldfield was required to wear an electronic tag for two months.[22] Then, having been denied a spousal visa to live in Britain and under the threat of deportation, Oldfield decided to appeal against the Home Office's decision, invoking Article 8 of the European convention on human rights, which guarantees the right to a family life, which he would be denied if deported to Australia.[23] Oldfield felt that it was "a very vindictive decision, political in nature, and an overreaction."[24]

Support of his appeal came from many quarters, including the universities and the rowing community, as well as others sharing his political views and protest ideals. For example, a letter/petition circulated and signed by 265 dons, students, and Oxbridge alumni[25] stated:

> The Boat Race is a game; its disruption should not result in any individual's deportation. Certainly its disruption should not be cause to separate an individual from his family, which includes a recently-born child.
>
> The race was completed successfully and no one was harmed by Mr. Oldfield's actions. We do not wish this draconian penalty to be applied in the name of an event representing our institutions.

His QC, Stephanie Harrison, described the threat of deportation as "grossly disproportionate." The appeal hearing was well attended by Oldfield supporters. Then, following appeal to the First-tier Tribunal (Immigration and Asylum Chamber), Oldfield was finally allowed to stay in the UK, when the appeal judge overturned the Home Secretary Theresa May's previous decision.[26] The judge, Kevin Moore, restated a view previously expressed by Judge Molyneux that Oldfield was a man of "character and commitment," adding there was no doubt that he was of value to UK society.

Although Oldfield's response to the judge's comments in summing up was emotional, he was specific in his intent, in the consistency of his beliefs, and the rationale for his protest. He offered his word that "we won't be here [in a tribunal] again." He insisted that his home was "here" ... in the UK, not in Australia, a "particularly racist country," on the other side of the world. Breaking down in tears, he repeated a previous comment that he had protested during the 2012 Boat Race because of a "feeling of heartbreak" at the growing inequality in British society.[27]

The Task

You've been asked to advise the Boat Race organizers about the broader issues of venue security and athlete and spectator safety at the annual event. However, some think that venue security is impossible for nearly 8 miles of river bank—so why bother?

Following discussions, you've decided that a good approach would be to make the organizers more aware of stakeholder needs and concerns; the possible impact on stakeholders of breaches in security; and the ramifications for the Boat Race, if the stakeholders were adversely affected because of poor decisions or inaction by the organizers.

You also recognize that if something goes wrong, a whole set of emotion-driven responses and criticisms could follow, creating a media storm. So, you need to be able to complete and communicate a thorough and relevant stakeholder analysis.

In addition, you realize that it would also be advantageous for the stakeholder analysis to be tied in with a competent risk analysis and linked to a risk and crisis management plan. To be more convincing, you also realize that it would be a good idea to draw parallels to similar protest and terror threats that have disrupted other sporting and political events and pass on any lessons to be learned.

The following section sets out the structure for the analysis that we have decided would ensure that no stone is left unturned or overlooked. So, go to it!

Study Questions

1. Why has the case attracted such media attention?
    a. Describe why the protest on the Thames created so much public interest and attracted such attention in the sports world.
    b. Comment on the social context, the athletes, the sport, the fans, and the media.
    c. Outline the events associated with the protest in chronological order, describe Oldfield's escalation of involvement, and identify moments when he could have made different decisions.
    d. Draw parallels with examples of protests that have targeted the operations of business organizations and political events.
2. Outline the major "players" involved in, or affected by, the protest on the Thames; that is, identify the key stakeholders relating to the Boat Race and the Boat Race incident:
    a. Describe their main attributes, responsibilities, stakes, and interests.
    b. Describe how the stakeholders impacted and were impacted by public perception of the sports event and the public protest.

c. Describe the moral emotions that appear to have surfaced in those commenting on events, and how those emotions may impact moral judgments and decisions taken by the different actors.

3. What measures should the Boat Race organizers take to combat the threat of public protest (1) to the operation of the Boat Race as a sporting event, (2) to athlete and spectator safety at the event, and (3) to the future of the Boat Race as an event?

    a. First, prepare an *event risk evaluation*. Outline their roles within the sport system, within commerce, within the political world, and so forth.

    b. Prepare a broader *risk management analysis* as a basis for a risk management plan.

    c. Develop a risk and crisis management plan. What would the major features of a risk management plan be for (1) the 2013 Boat Race or (2) future Boat Race events?

4. Was the punishment of Trenton Oldfield appropriate? Consider the nature of the punishment imposed upon Trenton Oldfield: Did the punishment fulfill natural, procedural, distributive, or compensatory justice?

5. Determine the relevance to business and political environments:

    a. What lessons would be relevant to the planning and conduct of:
        i. Other relevant mega-sporting events, such as the 2012 London Olympic Games
        ii. Other major business and political events, such as the AGM of a Fortune 500 company, or a G20 Heads of State meeting

    b. Outline other future trends and implications that may arise from this case.

> **Note**
>
> Students could use Badaracco's framework,[28] Freeman's Stakeholder Approach[29] to strategic management, and Mitchell et al.'s Stakeholder Typology[30] to identify stakeholders; their attributes—orientation, stakes, interests, opinions, and power; their responsibilities; the stakeholder dynamics; and so on.
>
> Students could also use the Ideology and Political Risk Analysis framework of Thoma & Chalip[31] derived from Allison[32] and Coplin & O'Leary.[33]

Endnotes

1. BNY, The BNY Mellon Boat Race, 2014, http://theboatrace.org/men/the-course.
2. CUBC, Boat Race History, 2014, http://www.cubc.org.uk/cubc-history/history/.
3. Anon, "New Sponsor for the Boat Race," 2012, http://www.sport.cam.ac.uk/news/new-sponsor-for-the-boat-race.html.
4. K. Cassidy, "Convicted Boat Race Protester Trenton Oldfield Fights Deportation from UK after Visa Denied," *ABC News*, November 28, 2013, http://www.abc.net.au/news/2013-11-28/trenton-oldfield-fights-deportation/5119444.
5. L. Davies and A. Bull, "Boat Race Rower Blasts Protester," *The Guardian*, April 8, 2012, http://www.theguardian.com/sport/2012/apr/08/boat-race-protest-reaction.
6. D. Aitkenhead, "Boat Race Protester Trenton Oldfield: 'I Had No Choice But to Swim,'" *The Guardian*, June 29, 2013, http://www.theguardian.com/world/2013/jun/29/trenton-oldfield-boat-race-protester.

7. ABC, "High Drama for Historic Oxford-Cambridge Boat Race," *ABC News*, April 9, 2012, http://www.abc.net.au/news/2012-04-08/protester-halts-historic-oxford-cambridge-boat-race/3938154.

8. D. Aitkenhead, "Boat Race Protester Trenton Oldfield: 'I Had No Choice But to Swim,'" *The Guardian*, June 29, 2013, http://www.theguardian.com/world/2013/jun/29/trenton-oldfield-boat-race-protester.

9. R. Booth, "Boat Race Protester Trenton Oldfield Wins Appeal Against Deportation," *The Guardian*, December 9, 2013, http://www.theguardian.com/world/2013/dec/09/boat-race-protester-trenton-oldfield-wins-appeal-deportation.

10. R. Myers, "Trenton Oldfield: The Joke's on Us with This Lunatic Deportation," *The Independent*, June 24, 2013, http://www.independent.co.uk/voices/comment/trenton-oldfield-the-jokes-on-us-with-this-lunatic-deportation-8671652.html.

11. ABC, "High Drama for Historic Oxford-Cambridge Boat Race," *ABC News*, April 9, 2012, http://www.abc.net.au/news/2012-04-08/protester-halts-historic-oxford-cambridge-boat-race/3938154.

    M. Cross, "Rowing Is Elitist, But Not in the Way Trenton Oldfield Thinks," *The Guardian*, April 9, 2012, http://www.theguardian.com/commentisfree/2012/apr/09/rowing-trenton-oldfield-boat-race.

12. L. Davies and A. Bull, "Boat Race Rower Blasts Protester," *The Guardian*, April 8, 2012, http://www.theguardian.com/sport/2012/apr/08/boat-race-protest-reaction.

13. L. Davies and A. Bull, "Boat Race Rower Blasts Protester," *The Guardian*, April 8, 2012, http://www.theguardian.com/sport/2012/apr/08/boat-race-protest-reaction.

14. P. McClean, "Boat Race Protester Found Guilty," *Cherwell*, September 26, 2012, http://www.cherwell.org/news/oxford/2012/09/26/boat-race-protester-found-guilty.

    M. Duell, "Boat Race Protester Trenton Oldfield Is Ordered Back Home to Australia as His Presence in Britain Is 'Not Conducive to the Public Good,'" *Daily Mail*, June 24, 2013, http://www.dailymail.co.uk/news/article-2347197/Trenton-Oldfield-ordered-leave-Britain-Home-.Office-jailed-disrupting-Boat-Race.html.

15. P. McClean, "Boat Race Protester Found Guilty," *Cherwell*, September 26, 2012, http://www.cherwell.org/news/oxford/2012/09/26/boat-race-protester-found-guilty.

16. L. Davies and A. Bull, "Boat Race Rower Blasts Protester," *The Guardian*, April 8, 2012, http://www.theguardian.com/sport/2012/apr/08/boat-race-protest-reaction.

17. L. Davies and A. Bull, "Boat Race Rower Blasts Protester," *The Guardian*, April 8, 2012, http://www.theguardian.com/sport/2012/apr/08/boat-race-protest-reaction.

18. K. Cassidy, "Convicted Boat Race Protester Trenton Oldfield Fights Deportation from UK after Visa Denied," *ABC News*, November 28, 2013, http://www.abc.net.au/news/2013-11-28/trenton-oldfield-fights-deportation/5119444.

19. M. Duell, "Boat Race Protester Trenton Oldfield Is Ordered Back Home to Australia as His Presence in Britain Is 'Not Conducive to the Public Good,'" *Daily Mail*, June 24, 2013, http://www.dailymail.co.uk/news/article-2347197/Trenton-Oldfield-ordered-leave-Britain-Home-.Office-jailed-disrupting-Boat-Race.html.

20. T. Garnett, "Trenton Oldfield Was Wrong to Sabotage the Boat Race. But Why Deport Him?" *The Guardian*, June 25, 2013, http://www.theguardian.com/commentisfree/2013/jun/25/trenton-oldfield-boat-race-why-deport.

H. Muir, "Boat Race Protester Trenton Oldfield Ordered to Leave UK," *The Guardian*, June 23, 2013, http://www.theguardian.com/world/2013/jun/23/boat-race-protester-trenton-oldfield-ordered-leave-uk.

J. Pettitt, "Oxbridge Dons Put Their Oar in over Boat Race Protester Trenton Oldfield's Deportation," *The Standard*, December 9, 2013, http://www.standard.co.uk/news/uk/oxbridge-dons-put-their-oar-in-over-boat-race-protester-trenton-oldfields-deportation-8992660.html.

21. H. Muir, "Boat Race Protester Trenton Oldfield Ordered to Leave UK," *The Guardian*, June 23, 2013, http://www.theguardian.com/world/2013/jun/23/boat-race-protester-trenton-oldfield-ordered-leave-uk.

22. M. Duell, "Boat Race Protester Trenton Oldfield Is Ordered Back Home to Australia as His Presence in Britain Is 'Not Conducive to the Public Good,'" *Daily Mail*, June 24, 2013, http://www.dailymail.co.uk/news/article-2347197/Trenton-Oldfield-ordered-leave-Britain-Home-.Office-jailed-disrupting-Boat-Race.html.

23. T. Garnett, "Trenton Oldfield Was Wrong to Sabotage the Boat Race. But Why Deport Him?" *The Guardian*, June 25, 2013, http://www.theguardian.com/commentisfree/2013/jun/25/trenton-oldfield-boat-race-why-deport.

24. H. Muir, "Boat Race Protester Trenton Oldfield Ordered to Leave UK," *The Guardian*, June 23, 2013, http://www.theguardian.com/world/2013/jun/23/boat-race-protester-trenton-oldfield-ordered-leave-uk.

25. Donsspeakout, "Oxbridge Speaks Out Against Oldfield Deportation Proceedings," *Donsspeakout*, December 8, 2013, https://donsspeakout.wordpress.com/2013/12/08/oxbridge-speaks-out-against-oldfield-deportation-proceedings/.

26. A. Bland, "Trenton Oldfield Deportation: Boat Race Protester Allowed to Stay in UK," *Daily Telegraph*, December 9, 2013, http://www.independent.co.uk/news/uk/politics/boat-race-protester-trenton-oldfield-wins-unprecedented-case-against-deportation-as-judge-overturns-theresa-mays-decision-8993322.html.

27. R. Booth, "Boat Race Protester Trenton Oldfield Wins Appeal Against Deportation," *The Guardian*, December 9, 2013, http://www.theguardian.com/world/2013/dec/09/boat-race-protester-trenton-oldfield-wins-appeal-deportation.

28. J. L. Badaracco Jr., *Defining Moments* (Boston, MA: Harvard Business School Press, 1997).

    J. L. Badaracco Jr., *Leading Quietly* (Boston, MA: Harvard Business School Press, 2002).

29. R. E. Freeman, *Strategic Management: A Stakeholder Approach* (Boston, MA: Pitman Publishing, 1984).

30. R. Mitchell, B. Agle, and D. Wood, (1997). "Towards a Theory of Stakeholder Identification and Salience: Defining the Principle of Who and What Really Counts," *Academy of Management Review* 22 (1997): 853–886.

31. J. E. Thoma and L. Chalip, "Hosting an International Event," in *Sport Governance in the Global Community* (Morgantown, WV: Fitness Inform Tech Inc., 1996).

32. G. T. Allison, *Essence of Decision: Explaining the Cuban Missile Crisis* (Boston, MA: Little, Brown, 1971).

33. W. D. Coplin and M. K. O'Leary, *Introduction to Political Risk Analysis* (Croton-on-Hudson, NY: Policy Studies Associates, 1983).

Additional Reading

B. R. Agle, R. K. Mitchell, and J. A. Sonnenfeld, "Who Matters to CEOs? An Investigation of Stakeholder Attributes and Salience, Corporate Performance, and CEO Values," *Academy of Management Journal* 42, no. 5 (October 1999): 507–526.

K. R. Andrews and D. K. David, eds. *Ethics in Practice: Managing the Moral Corporation* (Boston, MA: Harvard Business School Press, 1989).

A. A. Elias, "Towards a Shared System Model of Stakeholders in Environmental Conflict," *International Transactions of Operational Research* 15 (2008): 239–253.

T. Jones and A. Wicks, "Convergent Stakeholder Theory," *Academy of Management Review* 24, no. 2 (1999): 206–221.

A. L. Mendelow, (1987). "Stakeholder Analysis for Strategic Planning and Implementation," in *Strategic Planning and Management Handbook*, 2nd edition (New York: Van Nostrand Reinhold, 1987), 176–191.

Telegraph, "Boat Race Protester Trenton Oldfield Wins Bid to Stay in UK," *Daily Telegraph*, December 9, 2013, http://www.telegraph.co.uk/news/uknews/law-and-order/10506059/Boat-race-protester-Trenton-Oldfield-wins-bid-to-stay-in-UK.html.

M. L. Stone, "Sporting Protests," *The Guardian*, April 10, 2012, http://www.theguardian.com/sport/gallery/2012/apr/10/sporting-protests.

M. C. Suchman, "Managing Legitimacy: Strategic and Institutional Approaches," *Academy of Management Review* 20, no. 3 (1995): 571–610.

Exhibit

Exhibit 17.1 Sport as a Vehicle for Protest

Date	Sport	Protest Event
June 4, 1913	Horse racing	Suffragette Emily Davison threw herself under King George V's horse Anmer at the Epsom Derby.
November 22, 1969	Rugby	An anti-apartheid demonstrator entered the field of play at the Twickenham ground during the London Counties versus South Africa match Springboks Tour.
August 19, 1975	Cricket	The Headingley cricket pitch was vandalized by a "Free George Davis" campaigner—protesting a wrongful conviction—leading to the abandonment of a cricket Test Match.
July 25, 1981	Rugby	Anti-tour, anti-apartheid protesters entered the field of play at Waikato Stadium, Hamilton, New Zealand, to stop a match during the tour of South Africa's Springbok rugby union team. Following the cancellation of the match, the tour continued with further hostile protest against South Africa's system of apartheid/racial segregation.
July 2003	Motor racing	Catholic priest Neil Horan ran on to the track at the Silverstone British Grand Prix.
January 31, 2012	Football	John Foley handcuffed himself to the goal posts during the English Premier League match between Everton and Manchester City at Goodison Park, Liverpool.

Data source: Adapted from M. L. Stone, "Sporting Protests," *The Guardian*, April 10, 2012, http://www.theguardian.com/sport/gallery/2012/apr/10/sporting-protests.

Index

A

Accu-Med ADC, 113
ADCs (automated dispensing cabinets), 111
    benefit analysis, 114-116
    medications, 112-113
    stocking strategies, 113-114
advising Harlequins, 179
aftermath of
    Bloodgate Affair, 174-177
    Boat Race protests, 194
Allegheny County Port Authority, bus transit system, 153-155
alternative sources of energy, investing in energy efficiency measures, 35-37
American Freightways, 45
    rate quotes, LTL services, 59
analytics
    decision-making processes, 151
    parole boards, 147-150
        study questions, 150-151
Antique Haven, Inc.
    background of, 126-127
    change in shareholder's interest through sale of stock, 128-129
    termination of shareholder's interest, 127-128

A-P-A Transport, rate quotes (LTL services), 57
appeals
    disciplinary actions, Bloodgate Affair, 174-177
    to verdict in Bloodgate Affair, 174-177
Ascent Medical Solutions, 113
athletes, reaction to Boat Race protests, 195
automated dispensing cabinets. *See* ADCs (automated dispensing cabinets)

B

benefit analysis, ADCs (automated dispensing cabinets), 114-116
Bennett, Steve, 61
Blackett, Jeff, 176
Bloodgate Affair
    advising Harlequins, 179
    aftermath of, 171-172, 174-177
    disciplinary actions, 172
        response to verdict, 173
    ethical dilemmas, 181
    governance, 182
    incident, 170-171
    outcomes of, 177-178
    study questions, 180-182

Boat Race
    history of, 191-192
    judgments against protestor, 196-197
    outcomes of, 198-199
    protests, 192-193
        aftermath of, 194
        Oldfield, Trenton, 193-194
        reaction to, 195-198
        supporters of, 197-198
    study questions, 200-202
    tasks, 199-200
Boat Race Company Ltd. (BRCL), 191
Boğaziçi University, Saritepe Campus, energy efficiency measures, 30-32
Boğaziçi University Saritepe Campus Sustainable and Green Campus Initiative for Boğaziçi University, 29-30
Booth, Toby, 173
BRCL (Boat Race Company Ltd.), 191
Brennan, Steph, 172, 174
    disciplinary actions, 178
budgets, selecting EEM measures for, 32-33
buildings, energy usage. *See* energy usage in buildings
bus transit system, 153-155
    data for case analysis, 158-159
    demand behavior, 159
    models, 156-158
    pre-analysis, 155-156
    study questions, 159-161
business analytics, defined, xx
business plans, investing in energy efficiency measures, 37
business travelers, 85

C

C corporations, 123-124
Cambridge University, 191
Çamlibel, Emre, 31-32, 34
capacity availability, wind energy, 98-99
capital investment costs, wind energy, 98
Chapman, Dr. Wendy, 172, 175
    disciplinary actions, 177
Charleston Rigging, 131
    history of, 131-134
    marketing studies, 135-136
    opportunities, 134
Cheika, Michael, 171
classifying service failure, hotels, 90
clean energy, wind energy. *See* wind energy
Cleary, Mick, 173
closing of the books method, 124
CO_2 emissions, 31
collaborative planning, forecasting, and replenishment, 4
Consolidated Freightways, rate quotes (LTL services), 47
contracting LTL services, Hankey Industries, 45-46
contracts, selecting for LTL services, 48
corporations
    C corporations, 123-124
    S corporations, 124-125
        Antique Haven, Inc., 126-127
        change in shareholder's interest through sale of stock, 128-129

profit/loss, 124
termination of shareholder's interest, 127-128
costs
initial costs, replacement strategies, 141
operating and maintenance costs, replacement strategies, 142
wind energy, 98-99
CPFR (collaborative planning, forecasting, and replenishment), 4
Cross, Martin, 195
customer recovery, 91

D

data for case analysis, bus transit system, 158-159
Davenport, Thomas H., xx
Davenport Associates, 136
Davis, Susan, 69
decision-making processes, 151
declining cost of wind energy, 99-100
demand behavior, bus transit system, 159
demand forecasts, 3
TechnoMart, 4-9
VidoCo, 4-9
Depew, Jack, 126
developing strategies for product bundling, 66
disciplinary actions
Bloodgate Affair, 172
appeals, 174-177
response to verdict, 173
Brennan, Steph, 178
Chapman, Dr. Wendy, 177

Dong, Shi, 7
Driscoll, Larry, 69
drivers of costs, wind energy, 98-99
capacity availability and efficiency, 98-99
capital investment costs, 98
operating costs, 99

E

EDI (electronic data interchange), 4
EEM (energy efficiency measures), 30-32
investing in as alternative sources of energy, 35-37
offering as a service, 34-35
selecting for budgets, 32-33
selecting in multiple time periods, 33-34
efficiency, wind energy, 98-99
electricity, wind energy, 97-98
electronic data interchange, 4
energy,
natural gas, 100-102
wind energy. *See* wind energy
energy efficiency measures (EEM), 30-32
Boğaziçi University, Saritepe Campus, 30-32
investing in as alternative sources of energy, 35-37
offering as a service, 34-35
selecting for budgets, 32-33
selecting in multiple time periods, 33-34
Energy Service Companies (ESCOs), 34

energy usage in buildings, 27-29
　Boğaziçi University, Saritepe Campus, 29-30
Engeman, Roger, 77-80
equipment purchase decisions, Fayette China Company, 142-143
equipment replacement strategies
　Fayette China Company, 141-142
　salvage value, 142
ERC (European Rugby Cup Limited), 171
　appeals, 178
　disciplinary actions, 172
　　appeals, 174-177
ESCO business model, 34-35
ESCOs (Energy Service Companies), 34
ethical dilemmas, Bloodgate Affair, 181
European Rugby Cup Limited (ERC), 171
Evans, Mark, 174
Evans, Nick, 169-170, 173
EverClean Energy, Inc., 95
　background of, 96-97

F

Fang, Sheng, 6
Fayette China Company, 139
　background of, 140
　equipment purchase decisions, 142-143
　replacement strategies, 139, 141-142
fixed-charge problem, 156-158
forecast information, 4
forecasts, 103

G

Garrett, John, 193, 196
governance, Bloodgate Affair, 182
green initiatives, 17. *See also* sustainability initiatives
growth, India, 86

H

H. P. Davidson, 69
Hankey Industries, 43
　contracting for LTL services, 45-46
　logistics operations, 43-44
　selecting contracts for LTL services, 48
Hankey, Joshua, investing in energy efficiency measures, 43
Harlequins
　advising, 179
　disciplinary actions, 172
　　response to verdict, 173
　internal investigations, 174
　rugby, 169
Harris, Jeanne G., xx
Harrison, Stephanie, 199
Heineken Cup, 169
　Harlequins versus Leinster, 169
Herfindahl-Hirschman Index, 85
Hopley, Damian, 173, 176
hotels
　India, background of, 84-86
　minus occupancies, 83-84
　service failure, 86-88
　　classifying, 90
　service process improvement, 90
　service recovery, 89
　SM International, 83-84

hub subproject, 155
  bus transit system, 156
  models, 156-158
Hudspith, Karl, 195

I

India
  growth of, 86
  hotel industry, background of, 84-86
Industrial Technology, 69-70
  background of, 70-71
  intermodal routing options, 71-72
  selecting, routes for shipping, 72-73
initial costs, replacement strategies (Fayette China Company), 141
intermodal routing options, 71-72
investigations, into Harlequins, Bloodgate Affair, 174
investing in energy efficiency measures as alternative sources of energy, 35-37

J

James, Andrew, 95-96, 110
Jillings, Charles, 177
judgments against protestor, Boat Race, 196-197

K

Keyes, James, 132
Keyes, Robert, 131-134
kilns, replacement strategies (Fayette China Company), 141-142

L

labor, 79
lead time, 79
Leinster, rugby, 169
Lewis, Ken, 136
Li, Tao, 7
logistics operations, Hankey Industries, 43-44
Lolly's Restaurant, 77-80
  study questions, 80
LTL services
  contracting, Hankey Industries, 45-46
  rate quotes, 46-48
    Hankey Industries, 45-46
  selecting, contracts, 48

M

Mably, Prosecutor Louis, 196
MacArthur, Bartholomew, 140
MacArthur, Carl, 140
maintenance costs, replacement strategies (Fayette China Company), 142
market research, Point and Shoot Camera Shop (product bundling), 64-67
marketing studies, Charleston Rigging, 135-136
May, Home Secretary Theresa, 199
McDonell, Jason, 5-7
McKenzie, Rod, 175
media, reaction to Boat Race protests, 195
medications, ADCs (automated dispensing cabinets), 112-113

Mike's Product Packaging and Distribution (PPD) group, 17
minus occupancies, 83-84
    service failure, 86-88
Mitchell, Mitch, 197
models, bus transit system (hub subproject), 156-158
Molyneux, Judge Anne, 197
Moore, Brian, 171
Moore, Kevin, 199
multi-period settings, selecting EEM, 33-34
Myers, Rupert, 195

N

Naik, Deepa, 193
natural gas, 31-33, 100-102
    prices, 101
    variables
        affecting demand, 101
        affecting supply, 101-102
Nelson, David, 195
NGOs (nongovernmental organizations), sustainability initiatives, 16
Nike, 191

O

O'Connor, Roger, 171
officials of Boat Race, reaction to protests, 196
Oldfield, Trenton, 192
    Boat Race, 193-194
        aftermath of protests, 194
    reaction to Boat Race protests, 196
    verdict from trial about protest, 196-197

operating costs
    replacement strategies, Fayette China Company, 142
    wind energy, 99
opportunities, Charleston Rigging, 134
options for dealing with returned water systems shipping containers PWS (Pasadena Water Solutions), 22-25
organizers of Boat Race, reaction to, protests, 196
Otay, Emre N., 30, 32, 34
outcomes of Boat Race protests, 198-199
Oxbridge, 191
Oxford University, 191

P

P3 (People Planet Profit), 15-18
park-and-ride project, 154-155
    data for case analysis, 158-159
    study questions, 159-161
Parker Motor Freight, rate quotes (LTL services), 47-57
parole boards
    analytics, 147-150
        study questions, 150-151
    decision-making processes, 151
Pasadena Oil Equipment (POE), 18
Pasadena Water Solutions. *See* PWS (Pasadena Water Solutions)
People Planet Profit (P3), 15
Peretin, Cheryl, 61

pharmaceuticals, ADCs (automated dispensing cabinets), 111
   benefit analysis, 114-116
   medications, 112-113
   stocking strategies, 113-114
Pilger, John, 197
Pinsent, Sir Matthew, 192, 196
Pittsburgh, Allegheny County Port Authority, bus transit system. *See* bus transit system
POE (Pasadena Oil Equipment), 18
Point and Shoot Camera Shop, 61
   background of, 61-62
   market research, product bundling, 64-67
   product bundling, 62-64
      developing a strategy, 66
Port Authority bus transit system. *See* bus transit system
PPAs (power purchase agreements), 97
pre-analysis, bus transit system, 155-156
prices
   of natural gas, 101
   of wind energy, 98
product bundling, Point and Shoot Camera Shop, 62-64
   developing a strategy, 66
   market research, 64-67
profit/loss, S corporations, 124
protests, Boat Race, 192-193
   aftermath of, 194
   Oldfield, Trenton, 193-194
   reaction to, 195-198
   supporters of protester, 197-198

PWS (Pasadena Water Solutions), 15
   history of, 18-19
   sustainability initiatives, P3 (People Planet Profit), 15-18
   water systems shipping containers, 19-21
      options for dealing with returned containers, 22-25
      supply chains, 21-22

Q-R

Ran, Bo-Liu, 7
rate quotes, LTL services, 46-48
   Hankey Industries, 45-46
reaction to Boat Race protests, 195-198
regression, study questions (parole boards), 150-151
renewable portfolio standards (RPS), 97
replacement strategies, Fayette China Company, 139, 141-142
restaurants, Lolly's Restaurant
   expanding operations to include dinner, 77-80
   study questions, 80
Richards, Dean, 170, 172, 174
   appeals, 175-176
Richardson, Logan, 71
Ross, Robert, 139
routes for shipping, selecting (Industrial Technology), 72-73
RPS (renewable portfolio standards), 97
Rucinsky, Mary, 111

rugby
Bloodgate Affair
aftermath of, 171-172, 174-177
appeals, 174-177
disciplinary actions, 172
ethical dilemmas, 181
governance, 182
incident, 170-171
outcomes of, 177-178
response to verdict, 173
study questions, 180-182
Harlequins, advising, 179
Heineken Cup, 169
Harlequins versus Leinster, 169
history of, 167-168
RWC (Rugby World Cup), 168

S

S corporations, 124-125
Antique Haven, Inc., 126-127
change in shareholder's interest through sale of stock, 128-129
profit/loss, 124
termination of shareholder's interest, 127-128
salvage value, 142
Sanger, Michael, 43
Saritepe Campus, Boğaziçi University, energy efficiency measures, 30-32
Saritepe Campus, Boğaziçi University Sustainable and Green Campus Initiative for Boğaziçi University, 29-30
Scale, Lee, 149

SCC (social cost of carbon), 100
Searle, David, 196
selecting
contracts for LTL services, 48
routes for shipping, Industrial Technology, 72-73
service failure, hotels, 86-88
classifying, 90
service process improvement, hotels, 90
service providers, ESCOs (Energy Service Companies), 34-35
service recovery, hotels, 89
service recovery paradox, 91
services, offering EEM as, 34-35
shareholders, S corporations
change in shareholder's interest through sale of stock, 128-129
termination of shareholder's interest, 127-128
shipping, intermodal routing options, 71-72
shipping containers, water systems shipping containers
options for dealing with returned containers, 22-25
PWS (Pasadena Water Solutions), 19-21
supply chains, 21-22
shuttle subproject, 155
Simpson, Mary, 126
SM International, 83-84
service failure, 86-88
service process improvement, 90
service recovery, 89
Small Business Corporation Investment Act (1958), 124
social cost of carbon (SCC), 100

Southeast Ohio Women's Hospital (SOWH), ADCs (automated dispensing cabinets), 111-112
  benefit analysis, 114-116
  medications, 112-113
  stocking strategies, 113-114
Soyak Holding, 31
stakeholders, Bloodgate Affair analysis, 180
stocking strategies, ADCs (automated dispensing cabinets), 113-114
strategies, developing for product bundling, 66
study questions
  analytics, parole boards, 150-151
  Bloodgate Affair, 180-182
  Boat Race, 200-202
  bus transit system, 159-161
  Lolly's Restaurant, 80
Subchapter S corporation. *See* S corporations
supply chains, water systems shipping containers, 21-22
supporters of protester of Boat Race, 197-198
sustainability initiatives
  P3 (People Planet Profit), 15-18
  Sustainable and Green Campus Initiative for Boğaziçi University, 29
Sustainable and Green Campus Initiative for Boğaziçi University, 29
Sweat, Earl, 77-78

T

Tanner, Arthur, 171
tasks, Boat Race, 199-200
TechnoMart, 4-9
  demand forecasts, 4-9
termination of shareholder's interest, S corporations, 127-128
time periods, selecting EEM in multiple time periods, 33-34
Turkey, investing in energy efficiency measures as alternative sources of energy, 35-37

U-V

used machinery, salvage value, 142
variables
  affecting demand, natural gas, 101
  affecting supply, natural gas, 101-102
VidoCo, demand forecasts, 4-9

W

water systems shipping containers, 19-21
  options for dealing with returned containers, PWS (Pasadena Water Solutions), 22-25
  supply chains, 21-22
Williams, Tom, 170, 172-175

wind energy, 97-98
    declining cost of, 99-100
    drivers of costs, 98-99
        capacity availability and efficiency, 98-99
        capital investment costs, 98
        operating costs, 99
    EverClean Energy, Inc., 96-97
    PPAs (power purchase agreements), 97
    prices, 98
Wolfe, Mike, 15

X-Y-Z

Xinhua Electronics Company, 69-70
Zeng, William, 195